Generalized Continuum Mechanics

and Engineering Applications

Series Editor
Yves Rémond

Generalized Continuum Mechanics and Engineering Applications

Angela Madeo

First published 2015 in Great Britain and the United States by ISTE Press Ltd and Elsevier Ltd

ISTE Press Ltd
27-37 St George's Road
London SW19 4EU
UK

www.iste.co.uk

Elsevier Ltd
The Boulevard, Langford Lane
Kidlington, Oxford, OX5 1GB
UK

www.elsevier.com

Notices
Knowledge and best practice in this field are constantly changing. As new research and experience broaden our understanding, changes in research methods, professional practices, or medical treatment may become necessary.

Practitioners and researchers must always rely on their own experience and knowledge in evaluating and using any information, methods, compounds, or experiments described herein. In using such information or methods they should be mindful of their own safety and the safety of others, including parties for whom they have a professional responsibility.

To the fullest extent of the law, neither the Publisher nor the authors, contributors, or editors, assume any liability for any injury and/or damage to persons or property as a matter of products liability, negligence or otherwise, or from any use or operation of any methods, products, instructions, or ideas contained in the material herein.

For information on all our publications visit our website at http://store.elsevier.com/

British Library Cataloguing-in-Publication Data
A CIP record for this book is available from the British Library
Library of Congress Cataloging in Publication Data
A catalog record for this book is available from the Library of Congress
ISBN 978-1-78548- 032-4

Printed and bound in the UK and US

Contents

Acknowledgments

The idea behind this book came from the writing of my Habilitation Thesis that I defended in Lyon in December 2014.

As a matter of fact, this event was for me the trigger of multiple reflections concerning both my personal and professional life. Indeed, when we are faced with the realization of an achievement of the type mentioned above, very naturally and, let me say, abruptly, we are also brought to make a point on what was before and on who were the people contributing to make you the person that you actually are. I can count such persons on one hand, but despite the scareceness of such important encounters, what they left with me is, to my own scale of values, priceless.

The first person that I have to thank is, for sure, Francesco dell'Isola, my PhD supervisor and mentor for several years. I am grateful to him for a long series of reasons that would be unfair to summarize in a few lines. He initiated me into the scientific method and allowed me to see the problems of life with new awareness with respect to what I could do before. The clarity of thought, the way of approaching and solving difficult problems, the pleasure of understanding, the aptitude of looking at life as it is are all things that I feel I possess today and that I owe him to a big extent. It is thus more than natural for me of think of Francesco dell'Isola when a section of acknowledgments is being written, since I look at him as the predominant contributor to my current way of being and I feel that it is again time to thank him for that. And since when an achievement is realized we are brought to simultaneously think of the past and the future, I finally mention here my hope that future times will be as productive and enriching as the past ones have undoubtedly been.

Before starting my PhD, I spent a period of time at Virginia Tech for an exchange program which allowed me to obtain an American Master of Science degree. I feel the need to mention the important role that my advisor Norman Dowling played for the successful conclusion of such an experience. He and his wife helped me by letting me feel at home though I was so far from home that the only fact of quantifying the distance could have scared a 22 year-old girl. He mentored my studies in mechanical behavior of materials giving me the instruments for appreciating the power of a sane application of theoretical tools. I thank him for the fundamental contribution that he gave to my growth in that delicate period of my life.

In the last year of my PhD studies, I had the chance to meet Pierre Seppecher, who greatly contributed to the successful completion of my PhD Thesis. From him, I could learn how creativeness and originality can still be accompanied by humbleness with respect to science. The time that he spent explaining to me how to approach a complex problem in the clearest possible way is worth its weight in gold. I thank him for having taught me that "one needs to come back to his work one thousand times" before it can be considered to be fully metabolized and hence intelligible for others. Such lessons of science and life will accompany me for the future and my determination is now to become able to transform them into something concrete.

Arriving at INSA-Lyon as Associate Professor, I had hence the fortune to meet Philippe Boisse who was working on a class of materials which will be deeply discussed in one chapter of this book: that of fibrous composite reinforcements. It rapidly turned out while working with him that such microstructured materials could have taken advantage of a generalized continuum modeling by second gradient theories. I then had the intellectual satisfaction of finding a first application of the models that I had developed up to that point, actually becoming aware of their true potentialities and also of their limits. I thank Philippe for having helped me in pursuing my inclination toward science notwithstanding the external difficulties that often arise in everyday academic life. I will be forever grateful to him for such invaluable help in the search of a stable academic direction.

Among the people that I have to thank, a special place is reserved to those who are friends before covering any other specific role. Once again, they are very rare, but the evidence of their disinterested presence and support gives me the strength of looking at life with the awareness that we are somehow not alone in confronting it.

Thank you to my current and former PhD students Gabriele Barbagallo, Marco Valerio d'Agostino and Manuel Ferretti for their commitment to science and for their constant work also concerning the typesetting of figures and text in this book.

Last, but not least, I thank my father and mother, for having brought me up, loved me and accepted me as I am. Today that I am wise enough to understand the difficulties that they faced, I feel even more grateful to them and I do not need any other specific reason for doing so.

1

General Introductory Aspects

1.1. Introduction

The microstructure of materials is an essential feature for the design of engineering structures with improved performances. In these last decades, a huge effort has been made in the direction of conceiving new materials with specific microstructures for the sake of producing exotic mechanical behaviors both in the static and the dynamic regime. Such man-made artifacts, usually called metamaterials, indeed show peculiar material properties that cannot be found in natural materials and that can have multiple engineering applications [ENG 06, ZOU 09, ZHO 09, MAN 13, VAS 98, VAS 01, BOU 13].

It is conceivable, at the present stage of knowledge and technology, to direct a consistent scientific effort toward the conception of microstructured materials showing unusual behaviors which may be beneficial for the functioning of engineering structures and for their optimization. In fact, engineering structures designed using microstructured materials may show very interesting mechanical properties such as light weight, improved stiffness, easy forming processes and so on. Moreover, such materials could also be used for innovative applications in the field of vibration control and stealth technology. In fact, metamaterials are good candidates for the conception of wave screens and wave absorbers since they may show particular properties with respect to elastic and electromagnetic wave propagation.

It is thus understandable that the new concept of metamaterial is nowadays increasingly attracting the interest of physicists and mechanicians and that different microstructures are being conceived in order to obtain the desired macroscopic properties. Usually, metamaterials are obtained by suitably assembling multiple individual elements but arranged in periodic or quasi-periodic substructures in order to show exotic global mechanical behaviors. The particular shape, geometry, size, orientation and arrangement of their constituting elements can affect, for instance, the propagation of waves of light or sound in a manner not observed in natural materials, creating material properties which may give rise to unexpected engineering applications. Particularly promising in the design and description of metamaterials are those microstructures which present high contrasts in their mechanical properties: these microstructures, once homogenized, may produce generalized continuum media (see, for example, [PID 97, ALI 03, FOR 98, FOR 99a, FOR 02, KRU 98]).

Another way to conceive and produce metamaterials is that of optimizing their microstructures by means of statistical approaches (see, for example, [MAN 13] and references there cited). In this way, the obtained microstructures are not periodic anymore, but nevertheless they possess a statistical "hidden order" which allows the macroscopic material to exhibit very particular characteristics, especially for what concerns their behavior with respect to wave propagation. Such materials have been called hyperuniform and have the very interesting property of being isotropic at sufficiently large scales: their response to wave propagation does not depend on the direction of propagation of the considered wave. More particularly, the width of the band gaps which are observed experimentally does not depend on such a direction of propagation. This fact opens very interesting perspectives to the continuum modeling of these metamaterials. Indeed, an isotropic relaxed micromorphic model of the type presented in [GHI 13, MAD 13, MAD 14b, NEF 13] could be used for the macroscopic description of the onset of band gaps by introducing very few elastic parameters which could subsequently be fitted on the available experimental evidence.

1.1.1. *Mechanical models for metamaterials*

The main theoretical challenge related to the modeling of the mechanical behavior of metamaterials is the choice of the model which one wants to use. In fact, there are several possible approaches to the complex problem of considering the effect of microstructures on the overall mechanical behavior of real materials which basically belong to two philosophically distinct categories:

– start from the detailed description of the microscale to arrive to the description of the macroscale;

– start directly from the description of the macroscale somehow accounting for the presence of microscales.

We refrain here from a deep analysis of these two "philosophies", limiting ourselves to briefly discussing some of their advantages and disadvantages. Indeed, a remarkable literature exists based on the adoption of the first viewpoint: start from the microscopic properties of complex materials to arrive to the homogenized ones (bottom-up approaches, see, for example, [FRA 86, FOR 98, KRU 98, FOR 02, FOR 99a, PID 97, GRÜ 88, SEP 11]). From this respect, we can cite so-called homogenization models, multi-scale methods, upscaling procedures and so on. The common idea to all such approaches is to establish "*a priori*" the characteristics of the microstructure (e.g. topology, mechanical stiffnesses, distribution of different phases, etc.) and develop suitable tools to arrive at the global mechanical properties at higher scales. The main advantage of these methods is that they allow us to directly know how the macroscopic parameters are related to the microscopic ones. It is clear that such information is a really useful tool since it suffices to observe the characteristics of a given microstructure to arrive to the homogenized descriptors which can henceforth be used to describe the material behavior at higher scales. Nevertheless, some drawbacks can also be reported about such methods which are substantially related to the fact that a certain number of simplifying assumptions concerning the characteristics of the microstructure are usually needed and often become too restrictive to be able to give rise to a homogenized behavior which is fully representative of the real material behavior at higher scales. For example, some standard homogenization techniques intrinsically need the imposition of boundary conditions between representative cells and it is difficult to establish whether one type of boundary condition is more realistic than another. As a result, we can summarize by saying that it is true that the homogenized system keeps in its memory some peculiar informations about the microscopic characteristics of the system itself, but often the simplifying hypotheses which have been made at the level of the microstructure are too restrictive to assure that the obtained homogenized system is fully able to describe the real macroscopic material behavior.

The second possible type of approach is to start directly from the description of the macroscopic scale by developing models which are able to describe the average mechanical behavior of the considered microstructured materials by means of a relatively small set of macroscopic descriptors (top-down approach). The main advantage of this kind of approach is that real material behaviors can be described by means of few constitutive parameters at those macroscopic scales which are interesting from an engineering point of view. Moreover, the efficacy of the adopted macroscopic theory can be easily compared with experiments which can be conceived and reproduced on specimens having reasonable sizes to be handled without problems related, for example, to the smallness of the samples themselves. Finally, the real material behavior being described by a limited number of parameters, it is conceivable to design structures which have rather sophisticated shapes and large dimensions just relying on a few equations describing the global mechanical behavior of the considered structure. However, the drawbacks of such a type of procedure are twofold:

– one must know that, even if in a simplified macroscopic framework, the global theory must be complemented with some additional macroscopic descriptors if we want to model some macroscopic manifestations of the microstructure;

– it is often hard to accomplish the inverse task of relating the proposed macroscopic descriptors to precise characteristics of the microstructure.

Hence, we can conclude by saying that, if such macroscopic models are able to be more easily handled at scales which are particular to engineering design, some difficulties arise when one needs to precisely relate the used macroscopic descriptors to detailed microscopical properties.

In summary, at the current state of knowledge, there is no common agreement on which would be the correct approach to be used to model at best the mechanical behavior of metamaterials. Would a bottom-up approach be more consistent than a top-down? In other words, is it better to start from the characteristics of the single components of the microstructure and to obtain the homogenized properties, or conversely to try to get a simplified model with relatively few parameters which is somehow able to account for the macroscopic manifestation of the underlying presence of a microstructure inside the material? To our feeling, the answer is: it depends. If the scope is to control in detail how the microstructures affect the macroscopic behavior of the system, then a bottom-up approach seems to be mandatory. However, if with an averaged model we are able to describe the phenomena we are interested in, then there is no reason for not doing so. In the optic of dealing with big pieces of metamaterials in view of engineering design, it is not reasonable to propose the use of a model accounting for the single presence of all the constituents of the considered microstructures. A continuum model would possibly be the desirable choice.

In the framework of continuum theories, the systematic use of Cauchy theories may sometimes represent a too drastic simplification of reality, especially when dealing with metamaterials, since some essential characteristics related to the heterogeneity of microstructures are implicitly neglected in such models. Every material is actually heterogeneous if we consider sufficiently small scales: it suffices to go down to the molecular or atomic level to be aware of such heterogeneity. Nevertheless, very often, the effect of microstructure cannot be detected at the engineering scale. In such cases, continuum Cauchy theory is a suitable choice for modeling the mechanical behavior of considered materials in the simplest and more effective way. However, there are some cases in which the considered materials are heterogeneous even at relatively large scales and, as a result, the effect of microstructure on the overall mechanical behavior of the medium cannot be neglected. In such situations, Cauchy continuum theory may not be sufficient to fully describe the mechanical behavior of considered materials especially when considering particular loading and/or boundary conditions. It is in fact well known that such continuum theory is not able to catch significant phenomena related to concentrations of stress and strain or to specific deformation patterns in which high gradients of deformation occur and which are, in turn, connected to particular phenomena which take place at lower scales. Moreover, Cauchy models are not able to catch in an appropriate way the dynamical response of some microstructured materials showing dispersive behaviors or even frequency band gaps. Generalized continuum theories may be good candidates to model such microstructured materials in a more appropriate way (both in the static and dynamic regime) since they are able to account for the description of some macroscopic manifestations of the presence of microstructure in a rather simplified way.

We have to explicitly say that the heterogeneity of the microstructures alone is not sufficient to unveil the need of using generalized continuum theories against classical continuum ones. Indeed, anisotropic constitutive laws can bring a lot of information concerning the microstructures of considered materials even when remaining in the framework of classical continuum theories. For example, orthotropic constitutive laws can be useful for considering the fact that there are two preferred directions inside the material as a consequence of the fact that the components of the microstructure are oriented in some privileged patterns (as is the case, for example, for woven fibrous composite reinforcements). Nevertheless, such orthotropic constitutive laws are sometimes insufficient to account for some complex microstructure-related deformation patterns in which high gradients of deformation occur. If, as an example, we consider the case of woven fibrous reinforcements, we can easily convince ourselves that the local bending of the yarns is a microscopic deformation mechanism which has a concrete impact on the macroscopic behavior of the piece. Such local bending can be associated with a rapid variation of the shear angle between initially mutually orthogonal yarns which can be interpreted as a concentration of high gradients of shear deformation in thin transition layers. In order to describe such particular patterns in the framework of a continuum theory, second gradient or micromorphic theories must be used instead of classical first gradient ones. Hence, we can summarize by saying that the presence of microstructures can lead to different modeling needs at the level of macroscopic theories:

– the need for considering particular anisotropic constitutive laws which account for the fact that the underlying microstructure has a macroscopic effect on the material behavior for the simple fact of giving rise, for example, to privileged material directions or completely anisotropic behaviors. Such anisotropic constitutive laws can classically be integrated in standard Cauchy continuum models and are sensible to account for a wealthy of microstructure-related effects;

– the need for accounting for some specific behaviors which are usually associated with the description of microstructure-driven concentration of stress and strain inside the material. In such cases, generalized continuum theories may be of use for an improved modeling of the mechanical behavior of microstructured materials.

In the remainder of this chapter, we will present different specific problems in which generalized continuum theories actually bring important complementary information which is essential for a precise modeling of the behavior of the considered mechanical systems.

1.2. Generalized continuum theories and some possible applications

Generalized continuum theories naturally belong to the second of the categories mentioned at the beginning of section 1.1.1 (top-down approaches) and, in this chapter, we will try to analyze whether their possible use can provide some advantages when dealing with real engineering problems. We are of course aware that the first category previously discussed (bottom-up approaches) is as legitimate as the second one for approaching a wealthy of problems, but its study will not be the subject of the present book. Instead, we will focus on a discussion about the use of generalized continuum theories to model materials with microstructure: we regard such theories as a reasonable "engineering" compromise between the complexity of the model which we want to use and the detail at which microstructures can be described.

1.2.1. *Some basic aspects concerning generalized continuum theories*

In this section, we recall some very basic aspects concerning generalized continuum theories in order to let them also be accessible to the non-specialist readers. More precisely, we will make a point about some of the different existing generalized continuum models and we will try to point out which model can be useful to describe specific phenomena of engineering interest. Indeed, a vast literature exists concerning the development of "second gradient", "couple stress", "Cosserat", "micropolar", "micromorphic" models and so on which dates back to the works of the Cosserat brothers, Mindlin, Toupin, Germain, Eringen, Bleustein, etc. (see, for example, [MIN 64, MIN 68, MIN 65, BLE 67, COS 09, GER 73a, GER 73b, TOU 62, TOU 64, ERI 99, DEL 14c, DEL 14b]). Such generalized continuum theories are today experiencing a vehement revival since it is becoming more evident which are their potentialities concerning the macroscopic mechanical description of microstructured materials (see among many others [BOU 13, MAD 12, MAD 14b, DEL 09a, DEL 14a, FER 14, FOR 10, ASK 11, PLA 13, SCI 08, EXA 01, NEF 07, NEF 13, FOR 99b, FOR 01, LAK 82, NEF 06, YAN 82, YAN 81, PLA 14, RIN 14, ALT 13, AUF 13, PIE 09a, PIE 09b, ALT 10, ERE 14, ROS 13, YAN 10]). In this section we present and compare a class of such generalized theories and we highlight some of their possible applications which may be worth further study in view of technological innovation.

In order to review, in a concise way, the different possible types of generalized continuum theories which are usually encountered in the literature, we need to clarify that there are two main ways of generalizing classical Cauchy continuum theories, namely:

– keep the same kinematics as Cauchy theory (only the displacement field), but envisage more complicated expressions of the strain energy density letting it depend on higher gradients of such a displacement field;

– extend the kinematics of the considered medium (displacement field + supplementary degrees of freedom) and introduce suitable expressions of the strain energy density as a function of such a kinematical fields.

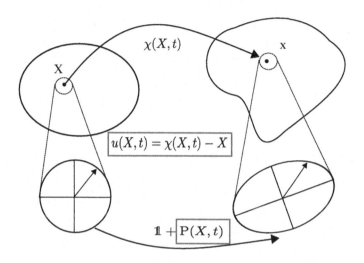

Figure 1.1. *Kinematics of a continuum with extra degrees of freedom*

In such a view of generalizing classical Cauchy continuum theory, we start by schematizing in Figure 1.1 the kinematics of a continuum with extra degrees of freedom. In addition to the classical placement field χ (or equivalently to the displacement u), we can generally introduce a second-order tensor field P which accounts for the deformation of microstructural elements embedded in the considered continuum. More precisely, if $\chi(X,t)$ highlights the current position of the macroscopic material particle X at time t, $P(X,t)$ can be seen as a deformation measure of the microstructures which are associated with the material particle X at the same instant. Depending on the complexity of the phenomenon that one wants to describe, more simplified generalized kinematics can be envisaged, by introducing fewer degrees of freedom with respect to the nine components of the tensor P.

Once the kinematical framework is established (only displacement field or displacement + extra degrees of freedom), the description of mechanical behavior of the considered generalized continuum can be achieved by the introduction of suitable constitutive laws for the strain energy density W. For what concerns generalized models with classical kinematics, we can cite some that are among the most known:

– *second gradient theories*: the strain energy density is a function of the first and second gradient of the displacement u, $W = W(\nabla u, \nabla\nabla u)$;

– *couple stress theories*: the dependence of the strain energy density on the higher order terms is made only through the gradient of the curl of displacement, $W = W(\nabla u, \nabla \mathrm{curl}\, u)$[1]. Of course, couple stress theories can be seen as a particular case of second gradient theories as explicitly pointed out, for example, in [MAD 15].

When considering generalized models with extended kinematics as the one presented in Figure 1.1, we can conceive a wealth of constitutive expressions for the strain energy density, so giving rise to different generalized models such as:

– *Classical micromorphic models*: the strain energy density takes the form $W = W(\nabla u, P, \nabla P)$, such models were originally considered by Mindlin and Eringen [MIN 64, ERI 99].

– *Relaxed micromorphic models*: the kinematics is exactly the same as in the classical micromorphic model, but the dependence on higher order terms is introduced only through the Curl of the microstrain tensor P. More precisely, the strain energy takes the form $W = W(\nabla u, P, \mathrm{Curl} P)$, where $(\mathrm{Curl}\, P)_{ij} = P_{ih,k}\epsilon_{jhk}$ and ϵ is the Levi-Civita alternator. Such a model was recently introduced in [NEF 13] for the isotropic case and, although the relaxed energy is not definite positive in the sense of Mindlin and Eringen, but only semi-definite positive, well posedness has been proven in [GHI 13].

– *Cosserat model*: only the skew symmetric part of the tensor P and its Curl appear in the energy, i.e. $W = W(\nabla u, \mathrm{Skew}\, P, \mathrm{Curl}\,(\mathrm{Skew}\, P))$. Since only the skew symmetric part of the microstrain tensor appears in the strain energy density, the actual kinematic unknowns can be reduced from 9 to 3.

– *Internal variable model*: the kinematics is the same as in Figure 1.1, but there is no dependence on the higher gradients of P in the strain energy density, i.e. $W = W(\nabla u, P)$. As we will better point out in the remainder of this chapter, such models may be useful to describe the mechanical behavior of systems with evolving microstructures.

It is worth noting that the generalized models with classical kinematics (second gradient and couple stress) can be seen as suitable particular limit cases of their more general counterparts with extended kinematics. More precisely, if we consider a classical micromorphic medium and let the microstrain tensor tend to the gradient of displacement, then a second gradient theory is obtained. Similarly, constraining a relaxed micromorphic model in such a way that the symmetric part of the microstrain tensor is vanishing, we get the Cosserat model as limit case (see also Figure 1.2). Even if this way of considering some generalized models as suitable limit cases of more general ones with extended kinematics is rather suggestive, the readers are warned that the two ways of approaching the same problem do not evidently coincide

---

1 The operator curl stands for the classical curl of a vector, i.e. $(\mathrm{curl}\, u)_i = \epsilon_{ijk} u_{k,j}$, where here and in the following we set $(\cdot)_{,j} = \partial\, (\cdot) / \partial X_j$ and ϵ_{ijk} is the Levi-Civita alternator.

when one is interested in the associated boundary conditions. Indeed, the internal and external actions that can be introduced in higher gradient theories are rather different from those which are particular to micromorphic media and the reduction of one set of boundary conditions to the other is not straightforward (see, for example, [BLE 67] for an example of reduction of micromorphic boundary conditions to second gradient ones in a simple linear-isotropic case).

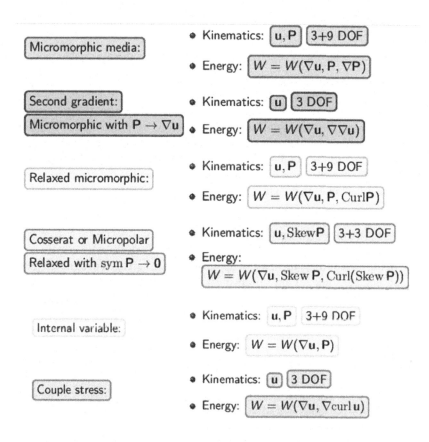

Figure 1.2. *Schematic presentation of different generalized continuum models*

Many scientific challenges related to the application of generalized continuum theories to the characterization and optimization of high-performance metamaterials can be identified. Nevertheless, each particular problem may need a different generalized model among the ones previously presented to have its best possible solution. We start to present and discuss some of these challenges and we will subsequently propose some specific strategies which can be conceived to overcome them by means of the exploitation of the most suitable generalized continuum model. In this chapter, we identify four main potential fields of applications of generalized continuum theories, which are likely to have important scientific and technological developments, namely:

– mechanical behavior of fibrous composite reinforcements (well described by second gradient models);

– wave propagation in metamaterials (well described by relaxed micromorphic and second gradient models);

– mechanical behavior of concrete (well described by internal variable models);

– mechanically driven remodeling of bone in the presence of bioresorbable materials (well described by internal variable models).

In this introductory chapter, we limit ourselves to proposing some general arguments supporting the idea that generalized continuum theories may be beneficial for the description of the quoted phenomena and we leave to the following chapters the task of giving a more precise and complete presentation of the main results which have been obtained in each domain.

1.3. Woven fibrous composite reinforcements

One of the most promising fields of application of generalized continuum theories is that of the modeling of the mechanical behavior of *woven fibrous composite reinforcements* (see, for example, [CHA 12, CHA 11, ORL 12]). Such metamaterials are constituted by two orders of yarns which have very high elongation stiffness, but very low shear stiffness. This strong contrast in the mechanical properties of the mesostructure is such that the homogenized material must necessarily be described at least in the framework of second gradient theories if we want to use a macroscopic continuum model. As a matter of fact, classical Cauchy theories are not sufficient for the description of specific deformation patterns usually observed in fibrous composite reinforcements such as concentration of high gradients of strains in thin transition layers which can be seen to be related to flexural strains of the fibers. It is worth stressing the fact that a classical Cauchy continuum theory is not able in any case to take into account the effect of flexural bending stiffness of the yarns on the overall mechanical behavior of fibrous composite reinforcements.

However, it is easy to understand that such a mesoscopic deformation mechanism may have an important macroscopic effect on the overall deformation of the considered material, at least for particular boundary conditions and/or applied external loads. In the remainder of this section, it will be pointed out how the bending stiffness of the yarns macroscopically affects the mechanical behavior of fibrous composite reinforcements at the engineering scale. It can be easily understood that such deformation mechanisms, which take place at lower scales, must be necessarily taken into account if one wants to fully characterize the behavior of fibrous composite reinforcements from a mechanical point of view. The macroscopic manifestation of mesostructure can indeed play an important role when considering the molding process of the reinforcement which may sometimes take complex shapes so allowing the conception of sophisticated engineering structural elements. Figure 1.3 shows, for example, a fiber-reinforced rotor blade which is conceived by molding the three-dimensional raw fiber reinforcement into the desired complex shape and then injecting a polymeric resin which confers the final stiffness to the engineering piece. It is clear that, during the forming process of the raw woven composite, the flexural rigidity of the yarns may play an important role in the final deformation of the blade. It is for this reason that a generalized continuum theory is a necessary step if we want to correctly predict the final deformed shape of the considered fiber reinforcements while remaining in a continuum framework.

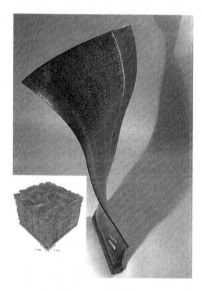

Figure 1.3. *Fiber reinforced rotor blade for use in an airplane engine (from [DEL 09c]). The detail shows an approximative weaving pattern of the 3D interlock constituting the blade (from [ORL 12])*

Analogous reasonings are also valid for composite materials which are constituted by fiber reinforcements immersed in a soft matrix: in this last case, generalized continuum theories are still useful even when considering the final composite material (woven fiber reinforcement + soft matrix).

1.4. Wave propagation in metamaterials

A second field of application of generalized continuum theories which may have a deep impact on engineering and technology is that of the study of *wave propagation in metamaterials*. In fact, classical Cauchy models are not sufficient to describe the dynamic behavior of metamaterials with complex microstructures. In fact, it is well known that real materials show dispersive behaviors especially when increasing the frequency of the traveling wave. This means that the dispersion relation (i.e. the relationship between the frequency ω of the traveling wave and the inverse of its wavelength λ) of a wave traveling in a real material cannot be considered to be a straight line. More complicated functional expressions $\omega = \omega(1/\lambda)$ are observed in real materials which significantly deviate from a straight line above all when considering sufficiently high frequencies. Such dispersive behavior cannot be caught in the framework of Cauchy continuum theories in which the dispersion relation is always linear. Generalized models, on the other hand, allow us to account for a more precise description of the dynamic behavior of materials since they are able to provide nonlinear dispersion relations. The dispersive behavior of a real non-dissipative material is more pronounced when considering waves with wavelengths which are small enough to interact with the microstructure of the material itself. For standard materials, such wavelengths often correspond to frequencies which are too high to be attained, so that a linear dispersion relationship still remains a good tool to predict the dynamic response of the considered medium from an engineering point of view. This is no longer true for engineering metamaterials in which the heterogeneities have characteristic sizes which may easily interact with waves at reasonable frequencies. For these reasons, generalized continuum models are seen to be useful to catch at best the dynamical properties of such metamaterials (see, for example, [CHE 12, BOU 13]).

Figure 1.4 shows two examples of metamaterials with periodic microstructures which may exhibit particular behaviors with respect to wave propagation in suitable frequency ranges.

The highly dispersive behavior of microstructured materials is only one of the interesting properties that they may show when considering wave propagation. Other characteristics which can render such metamaterials unexpectedly interesting for engineering applications are related to so-called frequency band gaps which are sometimes observed in architectured materials. It is indeed known (see, for example, [LIU 00, ZHO 09]) that some metamaterials obtained by suitably assembling periodic domains composed of materials with high contrasts of the mechanical properties may give rise to the onset of *frequency band gaps*, i.e. frequency ranges in which the macroscopic propagation of waves is inhibited. Such exotic dynamic behavior can be related to two main different phenomena occurring at the microlevel:

– *local resonance* phenomena (Mie resonance): the microstructural components, excited at particular frequencies, start oscillating independently of the matrix so capturing the energy of the propagating wave which remains confined at the level of the microstructure. Macroscopic wave propagation thus results in being inhibited;

– *diffusion* phenomena (Bragg scattering): when the propagating wave has wavelengths which are small enough to start interacting with the microstructure of the material, reflection and transmission phenomena occur at the level of the microstructure that globally result in an inhibited macroscopic wave propagation.

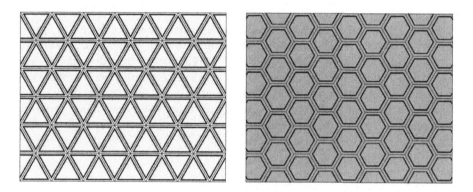

Figure 1.4. *Examples of periodic metamaterials*

Independently of the triggering mechanism, such inhibition of wave propagation typically intervenes for precise frequency ranges which are known as "frequency band gaps". Generalized continuum theories (relaxed micromorphic theories) may take into account the existence of such band gaps always remaining in the framework of a continuum theory. The advantage of using a continuum theory to describe this kind of phenomena may be found in the fact that a continuum model only provides a few elastic parameters to be tuned to fit experimental evidence. Moreover, numerical simulations based on suitably calibrated continuum models would allow a significant gain in terms of computation time with respect to discrete simulations explicitly accounting for the properties of the microstructures. The fact of adopting continuum models for describing the mechanical behavior of metamaterials may be somehow penalizing in the phase of conception of the metamaterial itself, since it is difficult to relate *a priori* the homogenized elastic parameters to the specific characteristics of the microstructure. However, a continuum model could be an incredibly powerful tool for the design of complex structures which are constituted by metamaterials with already known macroscopic generalized elastic properties. Indeed, the fact of dealing with a few macroscopic elastic coefficients makes the finite element implementation of the model rather easy, so allowing the possibility of designing complex structures with a relatively contained computational effort.

Other remarkable phenomena occurring at surfaces of discontinuities between two generalized continua may also be investigated. In fact, when considering the reflection and transmission of waves impacting at such surfaces, particular behaviors can be observed which are mostly related to the effect of "microscopic" boundary conditions on the amount of energy which is reflected and transmitted. In fact, a suitable choice of the boundary conditions to be imposed (at the miscroscopic level) between the two sides of the discontinuity surface can lead to complete transmission or reflection at the surface itself.

Finally, it is worth remarking that the study of the dynamic behavior of woven fibrous composite reinforcements in the framework of generalized continuum theories could also be of interest for further developments. Indeed, it has already been established that fibrous composite reinforcements are second gradient materials by considering the static case. This fact evokes the suggestion that the study of the dynamic behavior of such materials may unveil exotic behaviors which could be subsequently used for the conception of smart structures able to show peculiar characteristics with respect to wave propagation.

1.5. Reconstructed bone remodeling

Another possible field of application of generalized continuum theories may be found in the study of the biomechanical behavior of bone tissues in the presence or absence of bioresorbable artificial materials. It is indeed established both theoretically (see, for example, [WOL 86]) and experimentally (see, for example, [HAS 10, LAV 09]) in the scientific literature that the application of externally applied mechanical loads may favor the regeneration and remodeling of bone. It is clear that the application of suitable loads may be beneficial for the remodeling of bone also in the presence of bioresorbable materials. Among the generalized continuum models presented in section 1.2, we will show that an internal variable model is the most suitable one which can be used in order to describe the remodeling of a bone-biomaterial mixture in the framework of a continuum theory. Indeed, we will show in detail how a model of this type can be suitable to:

– account for the effect of the application of mechanical excitations on the macroscopic process of remodeling of both bone and biomaterial;

– include in the modeling the existence of an underlying activity of cells at the lower scales: even if the model used is intrinsically macroscopic, the evolution laws for the introduced internal variables allow us to account for the effect of cell activity on the density variation of both bone and biomaterial at the macroscopic scale.

We will analyze in detail how the introduced internal variable model permits us to catch some basic features of the process of reconstructed bone remodeling at the macroscopic scale.

We will also briefly sketch the idea that (see, for example, [BUE 03]) classical Cauchy-type continuum theories do not allow for the possibility of predicting scale effects in bone, when it is subjected to particular loading and/or boundary conditions. Such a behavior is related to the fact that bone is a hierarchically heterogeneous material, i.e. it can be considered as homogeneous at the scale of the millimeter, but it starts presenting heterogeneities at the scale of the micron.

Indeed, at this scale, quasi-periodic circular structures called osteons (see Figure 1.5) can be detected which confer highly heterogeneous properties to the material itself. It is for this reason that Cauchy continuum theory does not allow us to fully describe the behavior of bone at small scales when considering, for example, torsion-type loading conditions since the prediction made by the classical continuum theory starts being far away from the experimental evidence. However, when an enhanced Cosserat-type continuum theory is used (see [BUE 03]), the fitting with the experimental data becomes much more precise.

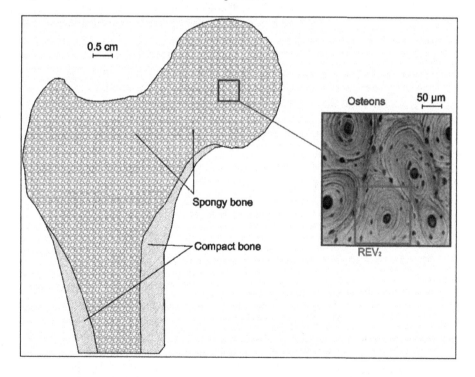

Figure 1.5. *First level of heterogeneity in natural spongy bone tissue*

If, on the one hand, it is accepted that the fact of using Cosserat, second gradient or micromorphic theories allows for a finer description of some microstructure-related deformation patterns in bone, on the other hand, the question which still remains open and which represents a huge scientific challenge is, instead, to establish whether and to what extent the effect of microstructure-induced deformations in bone can macroscopically affect the process of bone remodeling in living organisms. Even if second gradient or micromorphic models could bring some additional precision to the mechanical modeling of bone, above all for what concerns the possibility of describing deformation patterns in which high gradients of strain occur (for example, at the interface between the natural bone and artificial graft), we will limit ourselves here to introducing an internal variable model. This generalized model is in fact sufficient to catch the more basic biomechanical phenomena occurring in the process of reconstructed bone remodeling at the scale of the graft. We will show how a generalized continuum model with internal variable may be of use to describe the evolution of complex mechanical systems such as living bone tissue in the presence of resorbable grafts.

It is known today that precise biomechanical couplings exist in living tissues which allow us to use mechanical strains in the material to trigger biological activities which give rise to the so-called remodeling process (see, for example, [CAR 96, COW 76, COW 01, DOB 02, HUI 00, MUL 94, NOM 00, WOL 89]). Such a remodeling process basically consists of the resorption of bone tissue when the bone is too compact with respect to the small entity of the applied external load and, on the other hand, its synthesis when the applied external load is so high that more bone is necessary to correctly withstand the load itself. It is clear that the description of such biomechanical couplings is *"per se"* an enormous challenge for both mechanicians and biologists. We will focus ourselves on modeling such phenomena at the macroscopic scale and we will propose an internal variable model in which biomechanical couplings are accounted for and which is able to describe the evolution of both natural bone tissue and artificial bioresorbable bone substitutes as driven by externally applied mechanical solicitations.

1.6. Microstructure-driven energy dissipation in concrete

Another possible application of generalized continuum theories is the description of the *mechanical behavior of concrete*, with particular reference to energy dissipation in the dynamical regime. In fact, it is well known that concrete is an engineering material which can be considered to be almost homogeneous at the scale of the structure, but which indeed presents strong heterogeneities at lower scales due to the very different mechanical properties of its basic constituents. In particular, it can be supposed that the heterogeneous microstructure of concrete can be schematized by considering a set of microcracks immersed in a quasi-brittle matrix. Figure 1.6 shows the presence of the quoted microcracks in the cement matrix which confer to concrete particular mechanical properties related to the macroscopic effect of microscopic frictional sliding of the two superimposed sides of the microcracks.

The importance of considering the effect of the presence of microcracks on the overall mechanical behavior of such material is nowadays well recognized, and some authors propose to use rigorous homogenization techniques to derive the homogenized equations accounting for such heterogeneity (see, for example, [ZHU 08, AND 86, PEN 01]).

Generalized continuum theories are a valid tool for modeling the heterogeneity of concrete since they are able to account for the effect of the presence of microcracks on the macroscopic mechanical behavior of this material.

The most natural way of approaching the problem of modeling the heterogeneity of concrete in the framework of a generalized continuum theory is that of considering an enriched kinematics (in the spirit of Mindlin's and Eringen's micromorphic theories [MIN 64, ERI 99]) which is able to simultaneously account for:

– the standard macrodisplacement of the material points;

– the microscopic relative displacement of the crack lips.

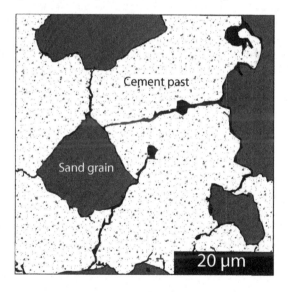

Figure 1.6. *Example of concrete microstructure:*
presence of microcracks

Once the correct macroscopical kinematical framework for the description of the complex motion of the considered microstructured material is established, then one can start conceiving constitutive relations for the strain energy density and, eventually, for the Hamilton–Rayleigh dissipation functional which are able to account for the deformation and dissipation phenomena which take place in concrete. The choice of such constitutive relationships is not trivial and some results are available in the literature which are issued from homogenization procedures providing suitable expressions for the macroscopic potentials accounting both for the macroscopic deformation of the matrix and the microscopic frictional sliding of the crack lips (see [ZHU 08]). Engineers and mechanicians should at this point merge their efforts to provide effective constitutive equations which are able to account for the prediction of realistic behaviors for what concerns both the macroscopic overall deformation of concrete and the effect of energy loss due to friction at the microscopic level. In the remainder of this book, we will support the idea that an internal variable model can be a good choice to model the dissipative behavior of concrete in the dynamic regime.

It is clear that other important aspects related to the mechanical behavior of concrete are of equal importance and thus deserve the appropriate attention. We refer, for example, to the evolution of damage especially when concrete is subjected to cyclic loads. Indeed, the quantity of cracks which are present in the concrete matrix is not constant, but evolve in time due to fatigue loading. Some authors have tried to rigorously approach such kinds of problems (see, for example, [MAR 85]), but a considerable effort still needs to be made in this direction. One possible way of approaching such problems has been found in the last decades in the idea of creating *fiber-reinforced concrete,* i.e. a material obtained by adding a fiber reinforcement to concrete with the aim of controlling cracking due to plastic shrinkage and drying shrinkage. The fact of adding fiber reinforcements to concrete improves some aspects related to cracking, but renders the topology of concrete meso- and microstructure far more complicated than the original one. Hence, it is likely that generalized continuum theories would also be of help to model, control and improve the mechanical behavior of such composite materials.

In the remainder of this book, we will point out how the generalized continuum model which we propose to use for the description of the mechanical behavior of concrete is also suitable for describing the behavior of some concrete-like materials which are obtained by adding suitable inert fillers to the cement past. We will indeed propose to use such generalized continuum model to conceive a metamaterial which is obtained by adding calcareous fillers to the cement past and which has the twofold merit of:

– dissipating more energy than standard concrete when subjected to cyclic loading;

– having an unchanged (or even improved) mechanical stiffness with respect to standard concrete.

Such a type of material would be useful for conceiving engineering structures which are close to sources of vibrations due to its improved dissipative capabilities (e.g. tramways, railways, etc.).

2

Fibrous Composite Reinforcements

In the engineering effort of designing new materials and structures, a constant demand is directed toward the search for better performances and new functionalities. A class of materials which is increasingly gaining attention is that of so-called metamaterials, for example materials exhibiting different mechanical responses at different scales due to different levels of heterogeneity. Indeed, the macroscopic mechanical behavior of such materials is influenced by the underlying microstructure, especially in the presence of particular loading and/or boundary conditions. Therefore, understanding the mechanics of meso- and microstructured materials is becoming a major issue in engineering. In this chapter we focus on a particular class of engineering metamaterials which are known as woven fibrous composite reinforcements.

Fiber-reinforced composites are generally characterized by high, specific strength and improved stiffness compared to other materials (see, for example, [DUM 87, LAD 85]). The fiber reinforcement confers desirable mechanical properties to the final engineering pieces including light weight, high stiffness in the direction of the reinforcing fibers and relatively easy forming processes. Such materials have broad, proven applications especially in the aircraft and aerospace sectors. Even more specifically, these materials are attractive for aircraft and aerospace structural parts such as rotor blades and other components. Moreover, there is also a growing interest nowadays in composite materials in other sectors such as the car industry. Additionally, fiber-reinforced composites have a decade-long history in military and government aerospace industries. The final engineering pieces can be obtained by molding the fibrous reinforcement into the desired complex shape which is finally maintained by injection and curing of a thermoset resin.

Fibrous composite reinforcements are constituted by woven tows which are themselves made up of thousands of fibers. Different weaving schemes can be used giving rise to different types of composite reinforcements (see Figure 2.1). While inside the unit cells of woven composites depicted in Figure 2.1 sharp changes in mechanical properties may occur, if the cells reproduce regularly, the homogenized material may be considered more or less homogeneous, since adjacent cells inside the material itself will globally have analogous properties. In fact, if we go down to the scale of the tows, the material can no longer be considered homogeneous, especially if the weft and warp have different properties.

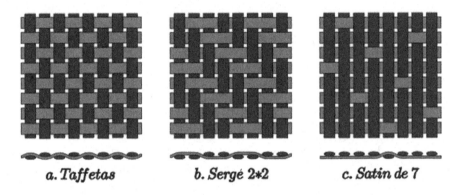

a. Taffetas   *b. Sergé 2*2*   *c. Satin de 7*

Figure 2.1. *Schemes of weaving for fibrous composite reinforcements*

All materials are actually heterogeneous if we consider sufficiently small scales, but woven composite reinforcements show their heterogeneity at scales which are significant from an engineering point of view. It is also clear that woven materials also macroscopically show strong anisotropy, since their mechanical response significantly varies if the load is applied in the direction of the fibers or in some other direction. Woven composite reinforcements can thus be considered to belong to the class of orthotropic materials, i.e. materials which have two privileged directions in their undeformed configuration.

As already pointed out, the fibrous composites sketched in Figure 2.1 can be easily shaped and their final shape is maintained by perfusing them with a hot organic resin which becomes rigid while cooling off. The final composite material commonly used in aerospace engineering is hence constituted by the fibrous composite material and the rigidified organic matrix. We are interested here only in describing the mechanical behavior of the fibrous composite reinforcements since this knowledge is fundamental for the process of formability of the final composite. Following [CHA 12], we find convenient to model the quoted fibrous reinforcements as continuous media. This hypothesis can be considered to be realistic if no relative displacement between fibers occurs. In other words, we assume that two superimposed yarns can rotate around their contact point, while no slipping takes place. This hypothesis is often verified during experimental analyses, even at finite strains. As it will be better pointed out in the remainder of this section, the anisotropy of the considered reinforcements will be taken into account by introducing suitable hyperelastic, orthotropic constitutive laws which are able to characterize the behavior of considered materials also at large strains.

A first gradient continuum, orthotropic model is not able to take into account all the possible effects that the microstructures of considered materials have on their macroscopic deformation. More precisely, some particular loading conditions, associated with particular types of boundary conditions may cause some microstructure-related deformation modes which are not fully taken into account by first gradient continuum theories. This is the case, for example, when observing some regions inside the considered fibrous reinforcements in which high gradients of deformation occur, concentrated in those relatively narrow regions which we will call transition layers. More generally, we can state that when flexural deformation modes take place at the scale of the yarns, such deformations immediately have a non-negligible macroscopic manifestation since they heavily influence the overall deformation of the considered macroscopic specimen.

The onset of shear boundary layers and localized flexural deformations of the yarns can be observed in some experimental tests which are used to characterize the mechanical properties of fibrous composite reinforcements. Indeed, in-plane local bending of the yarns does actually arise in the so-called bias extension test on two-dimensional (2D) woven composite reinforcements, while out-of plane bending of the yarns can be observed in the three point bending tests on thick three-dimensional (3D) composite interlocks.

2.1. Woven fibrous composite reinforcements modeled as second gradient materials

Fibrous composite reinforcements are materials with very special properties due to the hierarchical organization of their microstructure. Figure 2.2 shows an example of the different levels of microstructures which can be observed in such materials at different scales.

At the mesoscopic scale, fibrous reinforcements are very stiff in the warp and weft directions but have very low shear stiffness (i.e. the angle between yarns can vary rather easily). In this sense, such materials can be compared with some pantographic structures studied in [ALI 03, SEP 11] in which the authors present rigorous homogenization techniques to derive the macroscopic energies associated with the considered microstructures. Indeed, the pantographic structures presented in [ALI 03, SEP 11] are "idealized" structures in the sense that the authors suppose that the bars constituting the pantographs are infinitely rigid (no extension of the bars is permitted) and that angle variation between yarns may occur without associated energy. A deformation mode with no associated energy is called *floppy mode* by the authors. At this stage of the description, the presented structure could appear uninteresting since the bars are inextensible and there is no energy associated with the angle variation which can constitutively identify the structure. The system reduces to a structure in which an imposed angle variation in one point immediately propagates in all the material without the need of expending energy for doing so. In some sense, such pantographic structures closely resemble the mesostructure of fibrous composite reinforcements as shown in Figure 2.3. It is clear that the real composite will necessarily have some differences from the idealized structure since the yarns are not completely inextensible and some energy is needed to vary the angle between yarns due to, for example, friction phenomena or elastic interactions between the yarns themselves. Nevertheless, the presented schematization may be thought to be well adapted for the description of some basic deformation mechanisms intervening in 2D fibrous composite reinforcements.

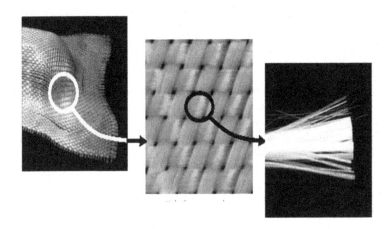

Figure 2.2. *Hierarchical organization of the microstructure in fibrous composite reinforcements. From left to right: macro, meso and micro scale from [CHA 11a]*

It is now interesting to observe that using the quoted simplifying assumptions for the considered structure (inextensibility of the bars and floppy mode for the angle variation) the authors of [ALI 03, SEP 11] are able to deduce that the homogenized macroscopic energy obtained from the quoted microstructure is a pure second gradient energy. This means that the mechanical behavior of the homogenized material is fully described by a deformation energy which only depends on the second gradient of displacement. In other words, the considered material is a pure and simple second gradient material and nothing else, since no first gradient energy is needed to describe its macroscopic deformation.

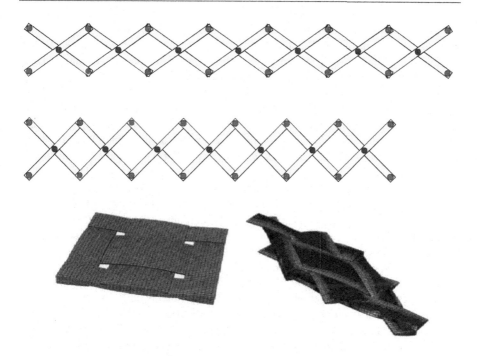

Figure 2.3. *Pantographic floppy mode of the structure presented in [ALI 03, SEP 11] (top) and deformation mode of the unit cell of considered fibrous composite reinforcements (bottom, [LOI 08])*

It must be noted that there is no actual way of describing a material as the one presented above in the framework of Cauchy continuum theories. Some more delicate modeling tools are needed which are particular to second gradient theories and which are basically related to:

– the fact that the macroscopic constitutive relation of the energy is given in terms of the second gradient of displacement (first gradient of strain);

– some more complicated boundary conditions with respect to the simple assignation of displacement or force are needed in second gradient theories in order to have a well-posed problem.

We have already dealt with the first point by means of the previous discussions in section 1.2.1 and such point will be more rigorously treated in the following. We now try to convince the reader that the second point is also necessary, i.e. non-standard boundary conditions must be imposed in order to unveil the "second gradient effects" on the deformation of the considered homogenized material. In doing so, we return to the concept of floppy mode and to the fact that the pantographic (micro-) structure described above is able to uniformly deform itself without storing any deformation energy and also very far (in the limit at infinite distance) from the point in which an angle variation is imposed. This fact is possible because we have not blocked all the possible floppy modes of the structure with the appropriated boundary conditions. If we block the displacement of one central pivot of the pantograph (as in the top Figure 2.3), this is still not enough since the propagation of deformation is not prohibited. In order to render the structure isostatic, we need to fix a second central pivot of the pantograph, for example the adjacent one. In this way, no supplementary floppy modes are possible for the considered (micro-)structure which is now able to resist to applied external solicitations with its own deformation pattern (which will basically consists of local flexural modes of the bars). At the level of the homogenized material, the imposed microscopic boundary condition is "translated" in the imposition of a vanishing normal derivative of the macroscopic displacement. Other boundary conditions are possibly imposable in the framework of second gradient theories, but we do not discuss them here, the basic scope of this chapter being to show the fact that a second gradient framework is generally well adapted for describing fibrous composite reinforcements.

In the remainder of this chapter, we will present all the technical details which are needed for the second gradient macroscopic description of fibrous composite reinforcements in the framework of a second gradient theory. We stress once again the fact that, the considered composites being more complicated than the idealized structure presented above, we found it convenient to introduce first gradient energies to account for some (small) extensibility of the yarns and for some (small) energy related to the angle variation between yarns. Nevertheless, in the light of the aforementioned discussion, the interest of considering theories including the inextensibility constraint (also at a macroscopic level) together with second gradient constitutive behaviors remains an important aspect which deserves to be studied more deeply in future researches. This issue opens interesting modeling and numerical questions since it is known that the numerical implementation of material behaviors showing very high contrasts in the elastic properties may produce sensitive numerical errors (see, for example, [HAM 13a, HAM 13b]).

2.2. Kinematics

In this section, we are interested in the introduction of the correct kinematical framework which is needed to describe the deformation of 3D interlocks in a macroscopic continuum framework. To do so, we follow the reasonings proposed in [DEL 14a] for the case of 2D networks in which suitable second gradient energies are proposed which account for the effect of yarns' bending stiffness on the deformation of the considered 2D woven fabrics. Fibrous composite interlocks are constituted by different layers of thin woven fabrics which are held together by a third weaving pattern. To account for the fact that such materials show two privileged material directions, we introduce two orthonormal vectors $\mathbf{D}_1$ and $\mathbf{D}_2$ which represent the warp and weft directions of the yarns constituting the 2D woven fabrics in the reference configuration. These weaving directions are the same for all points in the considered woven specimen. A third direction can be introduced as $\mathbf{D}_3 = \mathbf{D}_1 \times \mathbf{D}_2$: it is worth noting that while $\mathbf{D}_1$ and $\mathbf{D}_2$ actually identify the pattern of the yarns in the undeformed configuration, the third unit vector $\mathbf{D}_3$ does not necessarily represent a material direction. The quoted set of unit normal vectors is known to be worth to describe the reference configuration of an orthotropic, homogeneous material (see, for example, [RAO 09]). Once the Lagrangian unit vectors are introduced, we can define the corresponding Eulerian vectors as:

$$\mathbf{d}_1 = \mathbf{F} \cdot \mathbf{D}_1, \qquad \mathbf{d}_2 = \mathbf{F} \cdot \mathbf{D}_2, \qquad \mathbf{d}_3 = \mathbf{F} \cdot \mathbf{D}_3, \qquad [2.1]$$

where $\mathbf{F} = \nabla\chi$ is the gradient of the usual placement map χ. The vectors $\mathbf{d}_1$, $\mathbf{d}_2$ and $\mathbf{d}_3$ are the push-forward in the current configuration of the vectors $\mathbf{D}_1$, $\mathbf{D}_2$ and $\mathbf{D}_3$, respectively. It is worth to stress the fact that, while the vectors $\mathbf{d}_1$ and $\mathbf{d}_2$, locally represent the current directions of the warp and weft, the vector $\mathbf{d}_3$ cannot be related to privileged directions inside the considered orthotropic material.

We can summarize by saying that the kinematics of the considered continuum is univocally determined by the introduction of a suitably regular placement field $\chi : B_0 \rightarrow \mathbb{R}^3$ which maps the Lagrangian configuration $B_0 \subset \mathbb{R}^3$ of the considered body into the 3D Euclidean space. The deformation of the body is hence completely described by means of the deformation gradient $\mathbf{F} = \nabla\chi$ as in classical continuum mechanics. In this framework, if we introduce an orthonormal basis $\{\mathbf{D}_1, \mathbf{D}_2, \mathbf{D}_3\}$, the corresponding deformed vectors $\{\mathbf{d}_1, \mathbf{d}_2, \mathbf{d}_3\}$ are immediately found by means of equation [2.1]. The fact of identifying two of the Lagrangian material vectors (namely, $\mathbf{D}_1$ and $\mathbf{D}_2$) with the reference directions of yarns will be seen to be useful to describe, in an intuitive way, the deformation of the considered orthotropic material.

In fact, with reference to Figure 2.4 for the definition of the angles θ and γ, it is possible to remark that the shear strain S can be related to the total angle variation γ according to the formula:

$$S = \mathbf{d}_1 \cdot \mathbf{d}_2 = |\mathbf{d}_1| \, |\mathbf{d}_2| \cos(\theta) = |\mathbf{d}_1| \, |\mathbf{d}_2| \sin(\gamma), \qquad [2.2]$$

where $\gamma = \gamma_1 + \gamma_2$ is the total angle variation field between the two orders of yarns from the reference configuration to the current one and $|\cdot|$ represents the length of considered vectors. Analogously, $\lambda_1 = |\mathbf{d}_1|$ and $\lambda_2 = |\mathbf{d}_2|$ are a measure of the yarns' stretches: indeed, the elongations of the two orders of yarns with respect to the reference configuration can be easily obtained as $\lambda_1 - 1$ and $\lambda_2 - 1$, respectively.

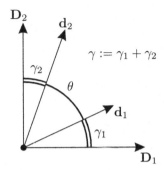

Figure 2.4. *Eulerian yarn vectors* d_1 *and* d_2*: the angle* θ *is the angle between yarns in the current configuration,* γ *is the total angle variation with respect to the reference configuration*

2.3. Second gradient energy density for 3D interlocks

The aim of this section is to introduce constitutive laws which are suitable to describe at best the mechanical behavior of 3D fibrous composite reinforcements. It will be shown that a second gradient constitutive law which is able to account for the out-of-plane bending stiffnesses of the yarns is really necessary to correctly model the mechanical behavior of such materials. Following what is done in [DEL 14a] for the 2D case, we suppose that the deformation energy density W depends on the deformation tensor and on its gradient by means of the following additive decomposition:

$$W\left(\mathbf{F}, \nabla\mathbf{F}\right) = W_I\left(\mathbf{F}\right) + W_{II}\left(\mathbf{F}, \nabla\mathbf{F}\right),$$

where W_I and W_{II} are the first and second gradient energies, respectively.

In order to determine a suitable constitutive expression for the first gradient energy W_I, we start recalling the representation theorem for orthotropic materials (see [RAO 09]) which states that the first gradient energy for an orthotropic material can take the following functional form:

$$W_I\left(\mathbf{F}\right) = W_I\left(i_1, i_4, i_6, i_8, i_9, i_{10}\right), \tag{2.3}$$

where

$$i_4 = \mathbf{D}_1 \cdot \mathbf{C} \cdot \mathbf{D}_1 = \lambda_1^2, \quad i_6 = \mathbf{D}_2 \cdot \mathbf{C} \cdot \mathbf{D}_2 = \lambda_2^2, \quad i_8 = \mathbf{D}_1 \cdot \mathbf{C} \cdot \mathbf{D}_2 = S,$$
$$i_9 = \mathbf{D}_1 \cdot \mathbf{C} \cdot \mathbf{D}_3, \quad i_{10} = \mathbf{D}_2 \cdot \mathbf{C} \cdot \mathbf{D}_3, \quad i_1 = \text{tr}\left(\mathbf{C}\right), \tag{2.4}$$

are the invariants of the right Cauchy-Green deformation tensor for an orthotropic material and where $\mathbf{C} = \mathbf{F}^T \cdot \mathbf{F}$ is the classical right Cauchy-Green deformation tensor. It is worth noticing that the first three invariants coincide, respectively, with the square of the stretches in the yarns' direction and with the shear strain. The invariants i_9 and i_{10}, on the other hand, are related to the out-of-plane angle variations of the two orders of yarns and their spatial gradient will be related to the out-of plane bending of the yarns.

As far as the second gradient energy is concerned, a general class of expressions which can be considered is of the type:

$$W_{II}\left(\mathbf{F}, \nabla \mathbf{F}\right) = W_{II}\left(\nabla i_1, \nabla i_4, \nabla i_6, \nabla i_8, \nabla i_9, \nabla i_{10}\right).$$

In the following, we will point out some reasonings which will allow us to consider simpler constitutive expressions for the second gradient energy which are suitable to describe the overall behavior of the considered interlock subjected to three point bending.

2.4. Constitutive choice for the first gradient energy

Following what is done in [CHA 12], we introduce some specific functions of the introduced invariants which are relatively simple to be determined by means of suitable experimental settings:

$$I_{\text{elong}}^1 = \ln\left(\sqrt{i_4}\right), \quad I_{\text{elong}}^2 = \ln\left(\sqrt{i_6}\right), \quad I_{\text{sh}}^p = \frac{i_8}{\sqrt{i_4 i_6}},$$

$$I_{\text{sh}}^{t1} = \frac{i_9}{\sqrt{i_4 i_{11}}}, \quad I_{\text{sh}}^{t2} = \frac{i_{10}}{\sqrt{i_6 i_{11}}}, \quad I_{\text{comp}} = \ln\left(\sqrt{\frac{i_3}{i_4 i_6}}\right), \qquad [2.5]$$

where the invariants which have not been previously introduced are defined as:

$$i_3 = \det\left(\mathbf{C}\right), \quad i_{11} = \mathbf{D}_3 \cdot \mathbf{C} \cdot \mathbf{D}_3. \qquad [2.6]$$

Indeed, considering an energy which depends on the quantities appearing in [2.5] is equivalent to consider a functional dependence of the type [2.3]. In fact, as shown in [RAO 09], the additional two invariants defined in [2.6] depend on the previously introduced ones by means of the following relationships:

$$i_{11} = i_1 - i_4 - i_6, \quad i_3 = \left(i_4 i_6 - i_8^2\right)\left(i_1 - i_4 - i_6\right) + 2 i_8 i_9 i_{10} - i_6 i_9^2 - i_4 i_{10}^2.$$

The interest of introducing a particular functional dependence of the strain energy density on the invariants [2.4] through the introduction of the quantities [2.5] can be found in the fact that these quantities can be easily measured by means of suitable experimental set-ups. The two quantities I^1_{elong} and I^2_{elong} are directly related to yarn elongations $\lambda_1 = \sqrt{i_4}$ and $\lambda_2 = \sqrt{i_6}$. As for the second quantity, it can be checked that $I^p_{sh} = sin\,(\gamma)$ (also see equation [2.2]): this means that it can be directly related to the shear angle variation between yarns. Analogously, I^{t1}_{sh} and I^{t2}_{sh} represent the out-of-plane angle variations of the two orders of yarns and are thus related to out-of-plane shear modes. Finally, I_{comp} represents a normalized volume variation which can be directly related to a compression deformation mode. The possibility of performing simple elementary measurements on the quantities [2.5] allows the conception of constitutive laws which characterize the behavior of composite interlocks and which show considerable agreement with the available experimental evidence. In [CHA 12], the authors propose a constitutive expression of the first gradient deformation energy of the type:

$$W_I = W^1_{\text{elong}} + W^2_{\text{elong}} + W_{\text{comp}} + W^p_{\text{sh}} + W^{t1}_{\text{sh}} + W^{t2}_{\text{sh}}, \qquad [2.7]$$

where

$$W^1_{\text{elong}} = \begin{cases} \frac{1}{2}\mathrm{K}^0_{\text{elong}}\left(I^1_{\text{elong}}\right)^2 & \text{if } I^1_{\text{elong}} \le I^0_{\text{elong}} \\ \frac{1}{2}\mathrm{K}_{\text{elong}}\left(I^1_{\text{elong}} - I^0_{\text{elong}}\right)^2 + \frac{1}{2}\mathrm{K}^0_{\text{elong}}I^I_{\text{elong}}I^0_{\text{elong}} & \text{if } I^1_{\text{elong}} > I^0_{\text{elong}}, \end{cases}$$

$$W^2_{\text{elong}} = \begin{cases} \frac{1}{2}\mathrm{K}_{\text{elong}}\left(I^2_{\text{elong}}\right)^2 & \text{if } I^2_{\text{elong}} \le I^0_{\text{elong}} \\ \frac{1}{2}\mathrm{K}^1_{\text{elong}}\left(I^2_{\text{elong}} - I^0_{\text{elong}}\right)^2 + \frac{1}{2}\mathrm{K}^0_{\text{elong}}I^{II}_{\text{elong}}I^0_{\text{elong}} & \text{if } I^2_{\text{elong}} > I^0_{\text{elong}}, \end{cases}$$

$$W_{\text{comp}} = \mathrm{K}_{\text{comp}}\left(\left(1 - \frac{I_{\text{comp}}}{I^0_{\text{comp}}}\right)^{-q} - q\,\frac{I_{\text{comp}}}{I^0_{\text{comp}}} - 1\right)$$

$$W^p_{\text{sh}} = \begin{cases} \mathrm{K}^{12}_{\text{shp}}\left(I^p_{\text{sh}}\right)^2 & \text{if } |I^p_{\text{sh}}| \le I^{p0}_{\text{sh}} \\ \mathrm{K}^{21}_{\text{shp}}\left(1 - |I^p_{\text{sh}}|\right)^{-p} + W^0_{\text{shp}} & \text{if } |I^p_{\text{sh}}| > I^{p0}_{\text{sh}} \end{cases} \qquad [2.8]$$

$$W^{t1}_{\text{sh}} = \begin{cases} \frac{1}{2}\mathrm{K}^{12}_{\text{sht1}}\left(I^{t1}_{\text{sh}}\right)^2 & \text{if } |I^{t1}_{\text{sh}}| \le I^0_{\text{sht1}} \\ \mathrm{K}^{22}_{\text{sht1}}\left(I^{t1}_{\text{sh}}\right)^2 + \mathrm{K}^{21}_{\text{sht1}}\left|I^{t1}_{\text{sh}}\right| + W^0_{\text{sht1}} & \text{if } |I^{t1}_{\text{sh}}| > I^0_{\text{sht1}} \end{cases}$$

$$W^{t2}_{\text{sh}} = \begin{cases} \frac{1}{2}\mathrm{K}^{12}_{\text{sht2}}\left(I^{t2}_{\text{sh}}\right)^2 & \text{if } |I^{t2}_{\text{sh}}| \le I^0_{\text{sht2}} \\ \mathrm{K}^{22}_{\text{sht2}}\left(I^{t2}_{\text{sh}}\right)^2 + \mathrm{K}^{21}_{\text{sht2}}\left|I^{t2}_{\text{sh}}\right| + W^0_{\text{sht1}} & \text{if } |I^{t2}_{\text{sh}}| > I^0_{\text{sht2}}. \end{cases}$$

In the previous formulas, all the quantities which have not been introduced before are constant.

We remark that expression [2.7] for the first strain energy density is not Legendre–Hadamard elliptic. This means that the solution of the Euler–Lagrange equations associated with this sole first gradient energy is not well posed and problems related to existence and uniqueness of the solution may thus arise. As a consequence, the fact that the considered first gradient problem is not well posed may lead to numerical simulations which are "mesh-dependent".

It is worth noting that the elongation energies W^1_{elong} and W^2_{elong} are defined in such a way that a threshold value I^0_{elong} exists for which the yarns' rigidity is smaller for small elongations than for higher ones ($K^0_{elong} < K_{elong}$). This constitutive choice allows us to take into account the fact that the yarns are not initially straight due to weaving and hence they can initially be elongated more easily. The elongation threshold I^0_{elong} corresponds to the configuration in which the yarns are completely straightened and start showing a higher resistance to deformation. The need of introducing such elongation strain energy densities is related to the fact that they actually carefully describe the response of the woven yarns to elongation. Nevertheless, the elongation of yarns is a mechanism which is definitely less important than the deformation mechanism associated with the angle variations between the two orders of yarns. We can actually say that, in most of the experimental tests, the considered yarns can be considered almost inextensible with respect to the observed predominant shear angle variation. As it will be better pointed out in the next section, this feature, which is peculiar of fibrous composite reinforcements, will be essential for choosing a simplified constitutive expression for the second gradient energy.

2.5. Constitutive choice for the second gradient energy

In this section, we specify the constitutive expression which will be used to model the mechanical behavior of 3D composite interlocks. To do so, we start considering some results recently proposed in [DEL 14a] for 2D woven composites. In the quoted paper, it is shown that a suitable 2D second gradient energy which is able to account for in-plane bending stiffness of the yarns at the mesoscopic scale is of the type:

$$W_{II}\left(\mathbf{F}, \nabla\mathbf{F}\right) = \frac{1}{2}A_\lambda \left(|\nabla\lambda_1|^2 + |\nabla\lambda_2|^2\right) + \frac{1}{2}A_S |\nabla S|^2,$$

where A_λ and A_S are positive constants and the elongation and shear strains have been, respectively, defined as $\lambda_1 = |\mathbf{F} \cdot \mathbf{D}_1|$, $\lambda_2 = |\mathbf{F} \cdot \mathbf{D}_2|$ and $S = \mathbf{D}_1 \cdot \mathbf{F}^T \cdot \mathbf{F} \cdot \mathbf{D}_2$. This energy has been shown to be a good choice for the description of the mechanical behavior of 2D woven composites due to its convexity with respect to $\nabla\mathbf{F}$ which guarantees well posedness of the resulting differential problem (see [DEL 14a]). A second gradient energy of this type has also been used in [FER 14] to model the bias extension test on 2D woven composites. It is clear that, when considering inextensible yarns, the gradients of elongations are vanishing and the second gradient strain energy density thus reduces to:

$$W_{II}\left(\mathbf{F}, \nabla\mathbf{F}\right) = \frac{1}{2}A_S\left|\nabla S\right|^2 = \frac{1}{2}A_S\left|\nabla i_8\right|^2.$$ [2.9]

At a careful observation, once chosen a suitable reference frame, such strain energy also closely resembles the one obtained in [ALI 03, SEP 11] by homogenization of a 2D pantographic truss. In [DEL 14a], it is also shown that, in the limit case of inextensible yarns, an alternative to the strain energy density [2.9] is given by:

$$W_{II}\left(\mathbf{F}, \nabla\mathbf{F}\right) = \frac{1}{2}A_g\left(\left|\mathbf{g}_1\right|^2 + \left|\mathbf{g}_2\right|^2\right),$$ [2.10]

where

$$\mathbf{g}_1 = \kappa_1\boldsymbol{\nu}_1, \qquad \mathbf{g}_2 = \kappa_2\boldsymbol{\nu}_2$$

are two vectors which account for the bending of the yarns at the mesoscopic level and A_g is a positive constant. In the last formulas, κ_1 and κ_2 are the bending strains of the two orders of yarns and $\boldsymbol{\nu}_1$ and $\boldsymbol{\nu}_2$ are the vectors orthogonal to the current yarn directions $\mathbf{d}_1$ and $\mathbf{d}_2$, respectively (see [DEL 14a] for more details). Direct comparison of equations [2.9] and [2.10] allows us to conclude that, in the case of almost inextensible yarns, the fact of considering an energy accounting for the gradient of the shear angle variation is equivalent to consider an energy accounting for the bending of the two orders of yarns at the mesoscopic level. This interpretation is intriguing since it provides a clear insight into the deformation mechanisms which take place at the mesoscopic level when considering woven fabrics.

By extension of the previous reasoning, we consider the following expression for the second gradient energy to be used to account for out-of plane bending stiffness of the yarns in 3D composite interlocks:

$$W_{II}\left(\mathbf{F}, \nabla\mathbf{F}\right) = \frac{1}{2}A_S^{t1}\left|\nabla i_9\right|^2 + \frac{1}{2}A_S^{t2}\left|\nabla i_{10}\right|^2.$$ [2.11]

By this constitutive choice, we are considering that the wires are almost inextensible (small elongations compared to the shear strains) and that the predominant second gradient deformation modes are the out-of-plane bending of the yarns at the mesoscopic level. This is coherent with the usual phenomenology observed when dealing with 3D composite interlocks.

As a matter of fact, the constitutive choice [2.11] for the second gradient strain energy density deserves more accurate investigations in future works in order to be generalized to describe any observable material behavior of thick composite interlocks. In fact, even if the predominant mesostructure-related deformation mechanisms which are activated in the three point bending test are the out-of-plane bending of the yarns, it is possible that other second gradient mechanisms could be activated when considering other loading and/or boundary conditions. In order to explore all these possibilities, other independent macroscopic tests need to be conceived which are able to unveil such supplementary material behaviors taking place at the mesoscopic level. A fully realistic constitutive choice for the generalized elasticity parameters remains a big challenge for mechanicians and it constitutes an open field of research. Despite the simplicity of the constitutive choice made here, nonlinear material behaviors are likely to occur also for second gradient deformation mechanisms. If equation [2.11] is well adapted for describing the macroscopic effect of the mesostructure when considering a macroscopic bending of the specimen, it is possible that more general expressions (including a dependence of the elastic second gradient parameters on the first gradient strain and/or more complicated functional expressions for the strain energy density) will be needed to describe the behavior of interlocks when subjected to arbitrary loading and boundary conditions.

2.6. Least action principle and principle of virtual powers

Once the kinematics and the adopted constitutive laws for 3D orthotropic materials have been introduced, we can introduce the action functional as:

$$\mathcal{A} = \int_{B_0} W\left(\mathbf{F}, \nabla \mathbf{F}\right) dB_0 = \int_{B_0} \left(W_I\left(\mathbf{F}\right) + W_{II}\left(\mathbf{F}, \nabla \mathbf{F}\right)\right) dB_0,$$

where W_I and W_{II} are constitutively given by [2.7] and [2.11], respectively. Assuming the previous expression for the action functional implies that all inertia effects are neglected and hence we are considering a static case. As will be shown in the remainder of the chapter, this assumption is sensible for the applications which are targeted here.

2.6.1. *Second gradient theory as the limit case of a micromorphic theory*

In this section, we present the principle of virtual powers for the considered second gradient material passing through the theory of micromorphic media. The theory of micromorphic media (see [MIN 64, ERI 99]) is known to be suitable to account for the presence of microstructure in elastic materials. This theory is more general than a second gradient one in the sense that the set of unknown kinematical fields is enriched with respect to the classical kinematics based on the displacement field alone (also see section 1.2). More precisely, supplementary kinematical fields accounting for the motion of the microstructure are provided thus generalizing the classical kinematical framework of Cauchy and second gradient continua. In this chapter, we state the principle of virtual powers for 3D composite interlocks by means of a simple micromorphic model and we use suitable Lagrange multipliers to let the considered micromorphic model "tend" to the second gradient model presented in the previous sections. The interest of introducing the principle of virtual powers via the micromorphic approach is threefold: (1) the presentation via a micromorphic model allows us to better catch the physical meaning of the considered internal and external actions, (2) the natural and kinematical boundary conditions which can be used naturally take an intuitive meaning and (3) last but not least, the numerical implementation of the considered generalized problem is easier and the obtained solution is more stable. As far as considering the third quoted advantage of using constrained micromorphic theories to numerically implement second gradient problems, we have to note that the gain in terms of numerical calculations is evident. In fact, when considering a differential problem arising from a micromorphic model, the associated differential equations are of lower order with respect to those which would directly derive from a second gradient model. In particular, the system of differential equations associated with a micromorphic model is of the second order, against the fourth order of that deriving from a second gradient theory. These lower order equations are obviously easier to be solved from a numerical point of view and the obtained numerical solution will be more stable and precise.

To proceed according to this optic, we introduce the kinematical fields of the considered micromorphic model by means of the two vector functions:

$$\chi : B_0 \to \mathbb{R}^3, \qquad \psi : B_0 \to \mathbb{R}^2,$$

the first one being the classical placement field introduced before also for the second gradient kinematics and the second one accounting for microscopic motions in the considered continuum. The micromorphic model proposed here is simpler than the classical one proposed by Mindlin and Eringen [MIN 64, ERI 99] since we only consider here two additional scalar functions instead of the nine which are introduced in the quoted models. Hence, we consider a micromorphic strain energy density which takes the following particular form and which is used to implement our numerical simulations:

$$\tilde{W}_{II}(\nabla\psi) = \frac{1}{2}A_S^{t1}|\nabla\psi_1|^2 + \frac{1}{2}A_S^{t2}|\nabla\psi_2|^2, \qquad\qquad [2.12]$$

where we denoted by ψ_α, $\alpha = 1, 2$ the components of the vector ψ. By direct comparison of the energies [2.12] and [2.11], it can be checked that the proposed micromorphic energy tends to the second gradient one introduced before if $\psi_1 \to i_9$ and $\psi_2 \to i_{10}$. In order to account for such constraints in the weak formulation of the problem, we introduce suitable Lagrange multipliers Λ_1 and Λ_2 which have an associated energy density of the type:

$$W_L\left(\mathbf{F}, \boldsymbol{\psi}, \boldsymbol{\Lambda}\right) = \Lambda_1\left(\psi_1 - i_9\right) + \Lambda_2\left(\psi_2 - i_{10}\right), \qquad [2.13]$$

where we clearly set $\boldsymbol{\Lambda} = (\Lambda_1, \Lambda_2)$.

Hence, we propose to write the action functional of the proposed micromorphic medium as:

$$\mathcal{A} = \int_{B_0} \left(W_I\left(\mathbf{F}\right) + \tilde{W}_{II}\left(\nabla\boldsymbol{\psi}\right) + W_L\left(\mathbf{F}, \boldsymbol{\psi}, \boldsymbol{\Lambda}\right)\right) \, dB_0,$$

where W_I is the same energy given in [2.7], while the energies $\tilde{W}_{II}$ and W_L are introduced in terms of the additional kinematical variables as in formulas [2.12] and [2.13], respectively. The power of internal forces of the considered constrained micromorphic medium can be written as the first variation of the considered action functional as:

$$\mathcal{P}^{int} = \delta\mathcal{A} = \int_{B_0} \left(\left(\frac{\partial W_I}{\partial F_{ij}} + \frac{\partial W_L}{\partial F_{ij}}\right)\delta F_{ij} + \frac{\partial W_L}{\partial \psi_\alpha}\delta\psi_\alpha\right.$$
$$\left. + \frac{\partial \tilde{W}_{II}}{\partial \psi_{\alpha,j}}\delta\psi_{\alpha,j} + \frac{\partial W_L}{\partial \Lambda_\alpha}\delta\Lambda_\alpha\right). \qquad [2.14]$$

The power of external forces is easily introduced when considering a micromorphic framework (see, for example, [BLE 67]) and in the present case, neglecting body external actions, can take the form:

$$\mathcal{P}^{ext} = \int_{\partial B_0} \left(f_i^{ext}\delta\chi_i + \tau_i\delta\psi_i\right). \qquad [2.15]$$

Indeed, in the performed numerical simulations for the three point bending we assume that the virtual field $\delta\psi_i$ is arbitrary on the boundary of the considered specimen (vanishing double force: $\tau_i = 0$), while the virtual displacement $\delta\chi_i$ is arbitrary almost everywhere, except on small subparts of the boundary where the displacement is assigned or vanishing. Such small parts of the boundary on which the displacement is vanishing can eventually change during deformation as happens for the contact of simply supported interlocks undergoing large bending deformations. The boundary conditions to be applied to model contact between two deformable continua are of difficult implementation but contact laws are usually already implemented in numerical codes as, e.g. COMSOL Multiphysics. We used such a tool to model the contact in our numerical three-point bending simulations.

The weak formulation of the differential problem for the considered constrained micromorphic medium can then be stated as:

$$\mathcal{P}^{int} = \mathcal{P}^{ext},$$

[2.16]

where the internal and external powers are, respectively, given by [2.14] and [2.15].

It is worth to remark that, starting from this formulation of the principle of virtual powers and considering arbitrary variations $\delta\Lambda_i$ of the Lagrange multipliers, we get the bulk constraints which actually let the considered micromorphic model tend to the particular second gradient one previously introduced, namely:

$$\psi_1 = i_9, \quad \psi_2 = i_{10}.$$

[2.17]

It is clear that, starting from the principle of virtual powers and integrating by parts, we could also obtain the strong form of the bulk equations and naturally associated boundary conditions in duality of the virtual variations $\delta\chi_i$ and $\delta\psi_i$. Nevertheless, since the numerical simulations presented in the following are directly implemented via the weak form [2.16], we do not explicitly write here such strong equations.

2.7. Numerical simulations for three point bending of composite interlocks

In this section, we present some numerical results arising from the application of the proposed second gradient model to the case of three point bending of a composite interlock. The first gradient constitutive parameters appearing in equation [2.8] are assumed to take the values presented in Tables 2.1, 2.2 and 2.3, in agreement with the experimental identification proposed in [CHA 12]. We remark that the two out-of-plane shear potentials are not symmetric in the sense that the corresponding constants appearing in Table 2.3 do not take the same values for the two order of yarns. This fact is due to different weaving patterns in the warp and weft directions and has been experimentally observed in [CHA 12].

K^0_{elong}	K_{elong}	I^0_{elong}	K_{comp}	I^0_{comp}	q
37.85 [MPa]	816.33 [MPa]	0.0145	7.57×10^{-3} [MPa]	-1.12	2.85

Table 2.1. Constitutive parameters appearing in the elongation and compression energy potentials

The physical test we want to reproduce here is a simple three point bending of a composite reinforcement beam with rectangular cross-sections. The considered interlocks are 3D materials (see Figure 2.5) in which a specific mesostructure with particular ordered patterns can be identified.

K_{shp}^{12}	K_{shp}^{21}	p	I_{shp}^{0}	W_{shp}^{0}
0.07575 [MPa]	1.69×10^{-4} [MPa]	3.69	4.2×10^{-3}	-1.69×10^{-4} [MPa]

Table 2.2. *In-plane shear constitutive parameters*

K_{sht1}^{12}	K_{sht1}^{22}	K_{sht1}^{21}	I_{sht1}^{0}
0.064945 [MPa]	0.00401131 [MPa]	0.00079691 [MPa]	1.4×10^{-2}

K_{sht2}^{12}	K_{sht2}^{22}	K_{sht2}^{21}	I_{sht2}^{0}
0.0330351 [MPa]	0.0042497 [MPa]	0.000736072 [MPa]	3×10^{-2}

Table 2.3. *Out-of-plane shear constitutive parameters*

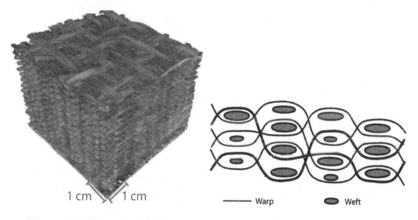

1 cm   1 cm   ——— Warp   ⬤ Weft

Figure 2.5. *Example of 3D woven composite (from [ORL 12]) interlock with an approximative weaving pattern*

As it can be inferred from Figure 2.5, such materials are realized by assembling different superimposed sheets of 2D woven fabrics by means of what is called a 2.5 weaving in the direction orthogonal to such sheets. For more details about the considered woven materials, we refer to [CHA 11, ORL 12] in which the mesostructures of Snecma composite interlocks are described in greater detail.

In this section, we will focus on two different types of samples which basically differ from each other for the relative orientation of warp and weft with respect to the boundaries of the macroscopic specimen. More particularly, we consider a three point $0°/90°$ bending test (warp and weft directions aligned with the edges of the specimen) and a three point $±45°$ bending test (the yarn directions form an angle of $45°$ with respect to the longer sample edge). Numerical simulations showing the effect of the introduced second gradient parameters will be proposed for both cases and a discussion on the need of considering such a generalized continuum theory will be performed.

In all the numerical simulations, we consider specimens with the dimensions of $200 \times 30 \times 15\,mm$, and we impose in the middle of the specimen a displacement of $60\,mm$. As already discussed, since large deformations are imposed to the specimen, the contact law between the specimen and the two cylindrical supports is a crucial point for the correct modeling of the considered problem. More particularly, as far as boundary conditions are concerned, we suppose that external double-forces are vanishing, while the forces at the supports are assumed to follow a frictionless contact law which is built-in in the COMSOL code. In the middle of the top surface, a displacement is applied which goes from 0 to $60\,mm$.

As for the values of the second gradient parameters appearing in equation [2.11], they are chosen according to Table 2.4 when not differently specified.

A_S^{t1}	A_S^{t2}
$[MPa \times m^2]$	$[MPa \times m^2]$
7.5×10^{-6}	7.5×10^{-6}

Table 2.4. *Values of the second gradient parameters used for the considered numerical simulations*

Such values have been chosen to fit the performed second gradient numerical simulations with the available experimental data concerning three-point bending of $0°/90°$ and $±45°$ composite interlocks. It must be noted that the quadratic second gradient constitutive model presented in equation [2.11] may not be sophisticated enough to catch all the deformation mechanisms which take place at the mesoscopic level. This insufficiency of the adopted constitutive second gradient model is suggested by the fact that the introduced elastic parameters seem to depend on the value of the imposed deformations and hence cannot be considered as real material constants. In order to overcome this problem, further studies on the formulation of the constitutive behavior of composite interlocks are needed which are focused on the development of more complex nonlinear second gradient constitutive laws. Moreover, the numerical stability of the searched solutions should also be studied in a deeper way in further investigations.

2.7.1. *Three point* $0°/90°$ *bending test: the effect of out-of plane yarns' bending stiffness*

In this section, we present the numerical simulations obtained via the proposed second gradient model and we compare the obtained solutions with those issued by the classical Cauchy theory.

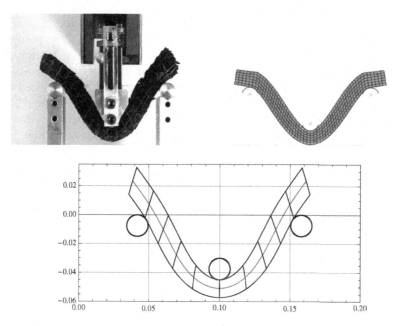

Figure 2.6. $0°/90°$ *three point bending for an imposed displacement of* $60 \, mm$. *left: experimental shape (from [ORL 12]), middle: first gradient numerical simulation and right: second gradient numerical simulation. For a color version of the figure see www.iste.co.uk/madeo/continuum.zip*

Figure 2.6 shows the comparison between the experimental tests, the first gradient solution previously obtained in [CHA 12] and the second gradient solution. It can be immediately noticed that the first gradient solution does not allow us to correctly describe the deformation of the two ends of the beam, since they do not lift up in the simulation which happens instead in the real experiment. Moreover, the curvature of the beam is not well described by the first gradient theory, while this problem is cured in the second gradient solution. The fact that in the first gradient solution no deformation of the two ends of the beam is observed is directly related to the fact that classical Cauchy continuum theory is not able to account for the yarn's bending rigidity.

The results obtained by means of the performed numerical simulations are appealing as they strongly suggest that the presence of second gradient terms in the strain energy density of the considered orthotropic material is unavoidable if we want to correctly model the three point bending of a $0°/90°$ interlock. In fact, it is sensible that the out-of-plane bending stiffness plays a very important role in the deformation of such materials. In fact, the predominant deformation mode in the considered test is related to the bending deformation of the order of yarns which is aligned with the longer side of the specimen. The order of yarns aligned with the depth of the specimen has very little influence on the global deformation of the considered sample. However, the fact that the longer yarns bend and that they posses a non-negligible out-of-plane bending stiffness allow the two ends of the beam to lift up. Such a deformation pattern is well recovered by the second gradient numerical simulations, but not by the first gradient ones (see Figure 2.6). It can be noticed that the cross-sections of the beam are not orthogonal to its mean axis both in the second gradient solution and in the real experiment: this is directly related to the fact that the yarns can be considered almost inextensible in the considered woven composite.

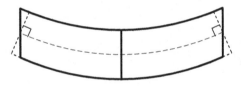

Figure 2.7. *Schematic representation of $0°/90°$ bending test: Euler–Bernouilli hypothesis of cross-sections orthogonal to the mean axis is violated*

Indeed, as it can be seen in Figure 2.7, a specimen which behaves as a Euler–Bernouilli beam (dash line) would need the upper part of the specimen to shrink and the lower part elongate in order to let the cross-sections stay orthogonal with respect to the mean axis. This shrinking/elongation deformation of the specimen is not possible due to the quasi-inextensibility of the yarns: the inextensibility constraint actually imposes a relative sliding of the yarns and, as a result, a rotation of the cross-sections with respect to the direction orthogonal to the mean axis.

2.7.2. *Three point* $\pm 45°$ *bending test*

The present section is devoted to the comparison between first and second gradient solutions for the $\pm 45°$ three point bending. Figure 2.8 shows a schematic representation of the bending of a $\pm 45°$ specimen.

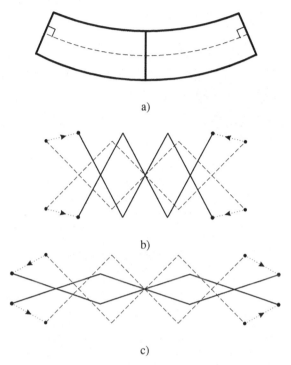

a)

b)

c)

Figure 2.8. *Schematic representation of a) ±45° bending test: Euler–Bernouilli hypothesis of cross-sections orthogonal to the mean axis is almost verified; b) and c) possible motions of pantographic structures allowing for elongation and shrinking of the macroscopic specimen*

It can be seen from this figure that in the deformation pattern shown in Figure 2.8(a), the upper part of the specimen necessarily undergoes shrinking, while the bottom part is instead elongated. This change of length of the specimen is not due to elongation of the yarns (which we know to be almost inextensible), but to their pantographic motions. As a matter of fact, it is known (see also Figures 2.8(b) and (c)) that pantographic structures can increase or decrease their global length without changing the length of the single elements constituting the pantograph itself. Such "pantographic" variation of length of the specimen, coupled to out-of-plane angle variation of the two order of yarns, gives rise to the overall deformed shape of the ±45° specimen.

Figure 2.9 shows the comparison between the experiments and the first and second gradient solution for the ±45° specimen for imposed displacements of 60 *mm*. It can be inferred from this figure that the first gradient solution is closer to the experimental shape than in the 0°/90° case. This means that the second gradient effects due to in-plane and out-of-plane bending of the yarns are less important than in the 0°/90°

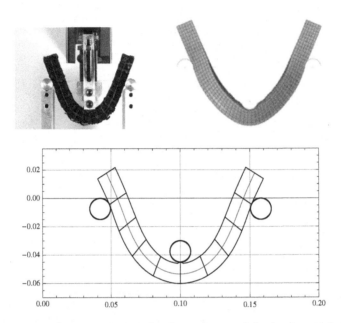

Figure 2.9. ±45° *three point bending for an imposed displacement of* 60 mm. *Left: experimental shape (from [ORL 12]), middle: first gradient numerical simulation and right: second gradient numerical simulation. For a color version of the figure see www.iste.co.uk/madeo/continuum.zip*

case. This is sensible since the yarns are short compared to the length of the specimen and hence they can deform (rotate) changing their out-of-plane shear angle with no significant bending. Nevertheless, the out-of-plane bending of the yarns still plays some role even if it is not so important as in the 0°/90° case. Such effect of the shear bending stiffness can be recognized to be important both for the complete lift of the two ends of the beam and for the curvature of the middle part of the specimen.

2.8. Bias extension test

In this section, we report some results which can be obtained by using a second gradient model similar to the one proposed for the 3D interlock to simulate a mechanical test on 2D sheets of woven composites which is known as bias extension test. We do not recall hear the precise constitutive choices made for the 2D case, referring to [FER 14] for additional details.

The bias extension test is a mechanical test which is very well known in the field of composite materials manufacturing (see, for example, [LEE 08, FER 14]). It is widely used to characterize the mechanical behavior of woven-fabric fibrous composite preforms undergoing large shear deformations. The bias extension test is performed on rectangular samples of woven composite reinforcements, with the height (in the loading direction) relatively greater (at least twice) than the width, and the yarns initially oriented at ± 45° with respect to the loading direction. The specimen is clamped at two ends, one of which is maintained fixed and the second one is displaced of a given amount. The relative displacement of the two ends of the specimen provokes angle variations between the warp and weft: the creation of three different regions A, B and C, in which the shear angle between fibers remains almost constant after deformation, can be detected (see Figure 2.10). In particular, the fibers in regions C remain undeformed, i.e. the angle between fibers remains at 45° also after deformation. However, the angle between fibers becomes much smaller than 45° in regions A and B, but it keeps almost constant in each of them.

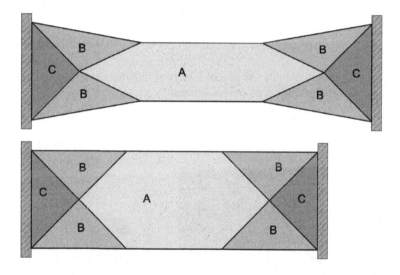

Figure 2.10. *Simplified description of the deformation pattern in the bias extension test*

The main characteristics of the bias extension test are summarized in Figure 2.10 in which both the undeformed and deformed shapes of the considered specimen are depicted. The specimen is clamped at its two ends using specific tools which impose the following boundary conditions:

– vanishing displacement at the bottom of the specimen;

– assigned displacement at the top of the specimen;

– fixed angle between the fibers (45°) at both the top and bottom of the specimen;

– vanishing elongation of the fibers at both the top and bottom of the specimen.

It is clear that the third type of boundary condition which imposes that the angle between fibers cannot vary during deformation of the specimen is a boundary condition which, at the level of a macromodel, imposes deformation and not displacement. The same is for the fourth type of boundary conditions blocking elongation of fibers. Boundary conditions of this type cannot be accounted for in a first gradient theory, while they can be naturally included in a second gradient one. Moreover, the deformation scheme described in Figure 2.10 does not take into account some specific aspects of the deformations which are actually observed during a bias extension test. In particular, the following two experimental evidences are not included in the scheme presented in the quoted figure:

– the presence of transition layers between two adjacent zones with constant shear deformation;

– the more or less pronounced curvature of the free boundaries of the specimen.

Indeed, both pieces of evidence can be observed in almost any bias extension test on woven composite preforms, as shown in Figure 2.11.

Figure 2.11. *Boundary layers between two regions at constant shear (left) and curvature of the free boundary (right). For a color version of the figure see www.iste.co.uk/madeo/continuum.zip*

A set of bias tests, performed on specimens under identical circumstances, has produced some suggestive results which were gathered in a picture of [LEE 08] which we reproduce here in Figure 2.12. In this figure, the contour of the shear angle variation between yarns is depicted as the result of some optical measurements conducted at INSA-Lyon.

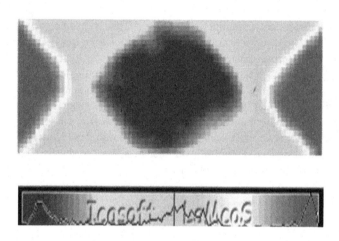

Figure 2.12. *Contour of shear angle in a bias extension test obtained from the optical measurement software Icasoft (INSA-Lyon [LEE 08]). For a color version of the figure see www.iste.co.uk/madeo/continuum.zip*

The principle of virtual powers for constrained 2D micromorphic media formulated by means of a suitable particularization of equation [2.16] allows for the description of the onset of thin boundary layers in which high gradients of shear deformation occur and which allow for a gradual transition from one value of the shear angle to the other one. The onset of these boundary layers cannot be accounted for by a first gradient theory, while it can be described by adding a dependence of the energy density on gradients of the shear deformation. Curvature effects will also be pointed out in the results obtained in the performed numerical simulations and which will be shown in the next section.

2.9. Numerical simulations

We now propose applying a 2D version of the introduced second gradient model (see [FER 14] for additional details) to perform numerical simulations of the bias extension test which take into account the onset of shear boundary layers. We consider a rectangular specimen of 100 mm of width and 300 mm of height in the undeformed configuration. The fibers are at $\pm 45°$ with respect to the direction of the height of the specimen in the undeformed configuration.

2.9.1. *First gradient limit solution*

As discussed in detail by [HAM 13a, HAM 13b], first gradient energies, in which the physical phenomena governing the onset of boundary layers are neglected, actually produce mesh-dependent numerical simulations. This is related to the fact that the proposed first gradient strain energy densities are not Legendre–Hadamard elliptic. In order to be able to prove existence of the solution for the hyperelastic problem, the notion of Legendre–Hadamard condition has been proved to be a useful tool, since it ensures ellipticity of the associated Euler–Lagrange equations (see, for example, [SCH 05, BIT 92]). It can be checked that the Euler–Lagrange equations associated with the only first gradient energy proposed here for woven composites are not L-H elliptic and this fact is strongly connected with the mesh dependence of the searched solution. As suggested in [DEL 14a], the fact of adding to the first gradient energy density a second gradient energy which is convex with respect to ∇F is useful to cure this inconvenience. In this way, the fact of using a second theory not only allows us to describe microstructure-related deformation phenomena, but also permits us to obtain a well-posed differential problem. To remedy the mesh-dependency of the searched solution, Spencer [SPE 84] suggested some techniques whose numerical counterpart has been developed in [HAM 13a, HAM 13b] for considered case: following the ideas there exposed, we could get numerical simulations in which boundary layers reduce to lines and deformation measures are subjected to jumps.

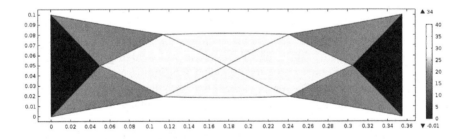

Figure 2.13. *Shear angle variation γ for an imposed displacement $d = 55\ mm$ obtained with the first gradient theory. The lateral bar indicates the values of γ in degrees. For a color version of the figure see www.iste.co.uk/madeo/continuum.zip*

We show the result of one of these numerical simulations in Figure 2.13. This picture represents the shear deformation field which is the correct limit to which regularized models must converge when higher gradient parameters tend to zero. In particular, Figure 2.13 shows the shear angle variation γ which is obtained as solution of the first gradient equilibrium problem. The boundary conditions which have been used to solve the first gradient equilibrium problem are:

– vanishing displacement on the left surface;

– assigned displacement on the right surface;

– unloaded lateral (top and bottom) surfaces (i.e. vanishing external forces).

As it can be seen, the three zones A, B and C defined in Figure 2.10 can be identified in the solution shown in Figure 2.13: the zones corresponding to regions C are such that no angle variation occurs with respect to the reference configuration ($\gamma = 0$). However, the zones correspondent to regions B and A, respectively are such that two different constant angle variations ($\gamma_B \approx \gamma_A/2 \neq 0$) with respect to the reference configuration occur. The first gradient solution is such that a sharp interface between each pair of the three shear regions can be observed.

2.9.2. *Second gradient solution and the onset of boundary layers*

For what concerns the detailed constitutive choice of the values of the needed second gradient parameters, we refer to [FER 14]. Further investigations are needed to establish a theoretical relationship between the microscopic structure of considered reinforcements and the macroscopic parameters here used it is actually well known (see, for example, [DE 81, FOR 10]) that the second gradient parameters are intrinsically related to a characteristic length L_c which is, in turn, associated with the microstructural properties of considered materials. It is also known that many identification methods have been introduced to relate the macroscopic second gradient parameters to the microscopic properties of the considered medium. Some of these methods are presented in [ALI 03, SEP 11] and must be the subject of future studies in order to relate the introduced macroscopic second gradient parameters to the microscopic bending stiffness of the yarns. Denoting by L_c the measured thickness of the shear boundary layer highlighted in Figure 2.11, we tune the value of the second gradient parameters in our numerical simulations until we obtain a boundary layer having the same thickness L_c.

The second gradient solution for the shear angle variation γ, obtained for the used values of the second gradient parameters, is shown in Figure 2.14. For obtaining this solution, the second gradient equilibrium equation was solved with the following additional (with respect to the first gradient case) boundary conditions:

– zero angle variation at the clamped ends of the specimen;

– zero elongation of the fibers at the clamped ends of the specimen;

– boundaries on the lateral (top and bottom) surfaces free from loads (vanishing external forces and double forces).

Such types of boundary conditions are not allowed in the framework of a classical Cauchy continuum theory and are peculiar of generalized continuum models.

It can be noticed that in the second gradient solution shown in Figure 2.14 the transition zones between different shear regions are regularized and shear boundary layers can be clearly observed, as well as a curvature of the free boundaries on the two free sides. It can be immediately remarked how the solution shown in Figure 2.14 is very close to the experimental picture shown in Figure 2.12.

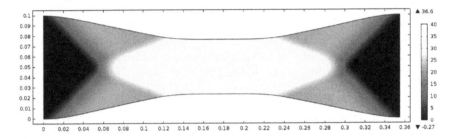

Figure 2.14. *Shear angle variation γ for an imposed displacement $d = 55\ mm$ obtained with the proposed second gradient theory. The lateral bar indicates the values of γ in degrees. For a color version of the figure see www.iste.co.uk/madeo/continuum.zip*

2.10. Conclusions

We have shown in this chapter that second gradient theories can actually be a useful tool for the accurate description of the deformation of 2D and 3D fibrous composite interlocks. We have highlighted different advantages that such a generalized continuum modeling can have in the respect to the description to the mechanical behavior of woven composite reinforcements, namely:

– second gradient theories can improve the modeling of fibrous reinforcements accounting for the effect of the bending stiffness of the yarns on macroscopic deformation modes;

– the adopted macroscopic theory can be implemented in finite element codes which can produce numerical simulations directly comparable with experiments that can be conceived and reproduced on specimens which have reasonable sizes to be handled without problems related, for example, to the smallness of the samples themselves. Such experiments thus serve as validation of the proposed model;

– real microstructure-related material behaviors can be described by means of few constitutive parameters at those macroscopic scales which are interesting from an engineering point of view;

– phenomena of stress and strain localization due to the presence of a highly heterogeneous mesostructure can be accounted for by such generalized continuum theories.

Notwithstanding the comforting results shown within this chapter, multiple further studies need to be developed in order to render the proposed model completely useful for the design and optimization of macroscopic engineering pieces, such as:

– set up suitable bottom-up procedures for relating the proposed macroscopic descriptors to the specific elastic properties of the microstructures;

– develop more general, non-linear second gradient constitutive laws for the strain energy density;

– set up explicit procedures for the standardized measurement of second gradient parameters.

We believe that the results presented in this chapter are a good example of a fruitful application of generalized continuum theories to problems of real engineering interest.

3

Wave Propagation in Generalized Continua

In recent years, the study of wave propagation in generalized continua has experienced a rapid development due to the possible advanced engineering applications that it may unveil. The author (and co-authors) of this book has been able to master a series of problems related to harmonic wave propagation, transmission and reflection in such media, so that a variety of different challenges seem to be clearly delineated. In particular, two big classes of problems may be identified, namely:

1) The study of the exotic behavior of generalized continua with respect to bulk wave propagation in infinite media (description of band gaps).

2) The problem of non-standard reflection and transmission of waves at discontinuity surfaces between two generalized media.

These two aspects regarding wave propagation are, of course, closely related, but they include at the same time some basic conceptual differences which will be pointed out in the following sections. Indeed, when we talk about the "exotic behavior" of a generalized continuum with respect to bulk wave propagation, we basically refer to its ability of describing so called *band gaps*, i.e. some frequency ranges corresponding to which waves cannot propagate inside the material. One of the mechanisms associated with such behavior is, once again, directly related to the fact that the considered generalized continuum possesses an underlying microstructure with which the propagating wave may interact. The physical mechanism which allows the inhibition of wave propagation is the following: if a signal is sent at a suitable frequency inside the considered metamaterial, it excites the microstructure which starts vibrating as well, creating local resonance phenomena that are able to trap the energy of the original wave which thus stops its propagation. Different metamaterials with precise topologies of microstructures have been conceived recently which effectively exhibit bad-gap phenomena (see

[LIU 00, VAS 98, VAS 01, ZHO 09]). We will show in the following sections that generalized continuum theories and, in particular, what we called "*relaxed micromorphic theories*", are able to account for the presence of such band gaps remaining in a macroscopic framework and with the need of very few constitutive parameters.

However, a second intriguing aspect concerning wave propagation in generalized continua can be investigated and is related to the modalities of reflection and transmission of waves in metamaterials. In fact, when considering two different generalized media which are in contact through a discontinuity surface, exotic reflection and transmission phenomena can be identified, even if the two considered metamaterials are not able to produce, per se, band gaps when considering bulk wave propagation (without discontinuity). In particular, it can be proven that intervening on (1) the material properties of the two metamaterials, (2) the boundary conditions to be imposed at the considered surface and (3) the material properties of the surface itself, one can act on the amount of energy which is reflected and transmitted also arriving to complete reflection or transmission if the introduced coefficients are suitably tuned.

In the following section, we show the results obtained both for the analysis of reflection and transmission of waves in generalized continua and the study of band gaps in relaxed micromorphic continua.

3.1. Band gaps in the relaxed micromorphic continuum

Micromorphic models were originally proposed by Mindlin [MIN 64] and Eringen [ERI 99] in order to study materials with microstructures while remaining in the simplified framework of macroscopic continuum theories. Nevertheless, the huge number of material parameters (18 in the linear-isotropic case) limited up to now the application of these powerful theories to describe the behavior of real metamaterials (see [NEF 07]). We propose to use the newly developed relaxed micromorphic model presented in [GHI 13, NEF 13] to study wave propagation in microstructured materials which exhibit frequency band gaps. The proposed relaxed model only counts six constitutive parameters and is fully able to account for the effect of microstructure on the macroscopic mechanical behavior of considered media. The request of positive-definiteness for the proposed relaxed model is weaker with respect to the classical Mindlin–Eringen model (positive-definiteness of the energy with the whole gradient term). This weaker request allows the relaxed model to account for weaker connections at the microscopic level compared to those which are possible in classical micromorphic theories. For this reason, the proposed relaxed model allows for the description of complete band gaps while the classical approach does not.

It is known that some materials like phononic crystals and lattice structures [VAS 01], granular assemblies with defects   [KAF 00, MER 10a, MER 11,

MER 10b] and composites [ECO 94, VAS 98, LIU 00] can inhibit wave propagation in particular frequency ranges (band gaps). We want to show here that the proposed relaxed model allows for describing frequency band gaps by "switching on" a unique constitutive parameter which is known as *Cosserat Couple modulus* μ_c (see [JEO 10, JEO 09, NEF 06, NEF 09, NEF 10]). The limited number of constitutive parameters makes the future conception of direct and indirect measurements on real materials exhibiting frequency band gaps possible. However, the generality of the proposed relaxed model can also be seen as a tool to aid the engineering design of new metamaterial-based structures with improved band gap capabilities. Materials of this type could be used as an alternative to piezoelectric materials which are used today for vibration control and which are for this reason extensively studied in the literature (see [AND 04, DEL 98, MAU 04, MAU 06, POR 05, VID 01]).

3.1.1. *Equations of motion*

As shown in [NEF 13], the equations of motion of the considered linear relaxed isotropic micromorphic continuum read:

$$\rho\,\mathbf{u}_{tt} = \mathrm{Div}\left[2\,\mu_e\,\mathrm{sym}\,(\nabla\mathbf{u} - \mathbf{P}) + \lambda_e\mathrm{tr}\,(\nabla\mathbf{u} - \mathbf{P})\,\mathbf{I} + 2\,\mu_c\,\mathrm{skew}\,(\nabla\mathbf{u} - \mathbf{P})\right]$$

[3.1]

$$\eta\,\mathbf{P}_{tt} = -\,\alpha_c\,\mathrm{Curl}\,(\mathrm{Curl}\,\mathbf{P}) + 2\,\mu_e\,\mathrm{sym}\,(\nabla\mathbf{u} - \mathbf{P}) + \lambda_e\mathrm{tr}\,(\nabla\mathbf{u} - \mathbf{P})\,\mathbf{I}$$
$$-\,2\,\mu_h\,\mathrm{sym}\,\mathbf{P} - \lambda_h\mathrm{tr}\,\mathbf{P}\,\mathbf{I} + 2\,\mu_c\,\mathrm{skew}\,(\nabla\mathbf{u} - \mathbf{P}),$$

where $\mathbf{u} \in \mathbb{R}^3$ is the displacement field, $\mathbf{P} \in \mathbb{R}^{3\times 3}$ is the microdeformation tensor (basic kinematical fields), ρ and η are the macro and micro mass densities, respectively, and all other quantities are the constitutive parameters of the model. As for the operators appearing in [3.1], we use standard notation. Such equations of motion can be seen to derive from the stationarity of the action functional associated with the Lagrangian $\mathcal{L} = T - W$, where the kinetic and potential energy are respectively given by:

$$T = \frac{1}{2}\rho\,\|\,\mathbf{u}_t\,\|^2 + \frac{1}{2}\eta\,\|\,\mathbf{P}_t\,\|^2,$$

[3.2]

and

$$W = \mu_e\,\|\,\mathrm{sym}\,(\nabla\mathbf{u} - \mathbf{P})\,\|^2 + \frac{\lambda_e}{2}\,(\mathrm{tr}\,(\nabla\mathbf{u} - \mathbf{P}))^2$$
$$+\,\mu_h\,\|\,\mathrm{sym}\,\mathbf{P}\,\|^2 + \frac{\lambda_h}{2}\,(\mathrm{tr}\,\mathbf{P})^2$$
$$+\,\mu_c\,\|\,\mathrm{skew}\,(\nabla\mathbf{u} - \mathbf{P})\,\|^2 + \frac{\alpha_c}{2}\,\|\,\mathrm{Curl}\,\mathbf{P}\,\|^2,$$

[3.3]

It can be checked that when considering a completely one-dimensional (1D) case, the term $\mathrm{Curl}\,(\mathrm{Curl}\,\mathbf{P})$ appearing in equation [3.1] vanishes and no characteristic length related to microstructure can be accounted for by our model. We need at least the case of plane waves (all the components of $\mathbf{u}$ and $\mathbf{P}$ are non-vanishing, but all depend on one space variable X which is also the direction of propagation of considered wave) to disclose all the characteristic features of the proposed relaxed model. However, Mindlin–Eringen models allow us to account for characteristic lengths even when considering complete 1D cases (all components of the kinematical fields in the plane orthogonal to the direction of propagation are vanishing). This is shown in [BER 09, MAU 99] in which these fully 1D equations are derived by the standard internal variable theory. We also remark that, in general, the relaxed term $\mathrm{Curl}\,(\mathrm{Curl}\,\mathbf{P})$ in the second of equations [3.1] is much weaker than the full term $\Delta\mathbf{P}$ appearing in Mindlin and Eringen models. Despite this weaker formulation, we claim that the proposed relaxed model is fully able to account for the presence of microstructure on the overall mechanical behavior of considered continua. In particular, our relaxed model is able to account for the description of frequency band gaps, while the classical Mindlin- and Eringen-type models are not.

In [NEF 13], it is also proved that positive definiteness of the strain energy density associated with equations [3.1] implies

$$\mu_e > 0, \quad \mu_c > 0, \quad 3\lambda_e + 2\mu_e > 0, \quad \mu_h > 0, \quad 3\lambda_h + 2\mu_h > 0, \quad \alpha_c > 0. \quad [3.4]$$

We limit ourselves to the case of plane waves traveling in an infinite domain, i.e. we suppose that the space dependence of all the introduced kinematical fields is limited only to the space component X which we also suppose to be the direction of propagation of the considered wave. We introduce the new variables

$$P^S := \frac{1}{3}\left(P_{11} + P_{22} + P_{33}\right), \quad P^D := (\mathrm{dev}\,\mathrm{sym}\,\mathbf{P})_{11}, \quad P_{(1\xi)} = (\mathrm{sym}\,\mathbf{P})_{1\xi},$$

$$P_{[1\xi]} = (\mathrm{skew}\,\mathbf{P})_{1\xi}, \quad \xi = 2, 3. \qquad [3.5]$$

It is immediate that, according to the Cartan-Lie-algebra decomposition (see [NEF 13]), the component P_{11} of the tensor $\mathbf{P}$ can be rewritten as $P_{11} = P^D + P^S$. We also define the additional variables

$$P^V = P_{22} - P_{33}, \qquad P_{(1\xi)} = (\mathrm{sym}\,\mathbf{P})_{1\xi},$$

$$P_{[1\xi]} = (\mathrm{skew}\,\mathbf{P})_{1\xi}, \quad \xi = 2, 3. \qquad [3.6]$$

We rewrite the equations of motion [3.1] in terms of the new variables [3.5], [3.6] and, of course, of the displacement variables. Before doing so, we introduce the quantities[1]

$$c_m = \sqrt{\frac{\alpha_c}{\eta}}, \qquad c_s = \sqrt{\frac{\mu_e + \mu_c}{\rho}}, \qquad c_p = \sqrt{\frac{\lambda_e + 2\mu_e}{\rho}},$$

[3.7]

$$\omega_s = \sqrt{\frac{2(\mu_e + \mu_h)}{\eta}}, \qquad \omega_p = \sqrt{\frac{(3\lambda_e + 2\mu_e) + (3\lambda_h + 2\mu_h)}{\eta}},$$

$$\omega_r = \sqrt{\frac{2\mu_c}{\eta}}, \qquad \omega_l = \sqrt{\frac{\lambda_h + 2\mu_h}{\eta}}, \qquad \omega_t = \sqrt{\frac{\mu_h}{\eta}}.$$

With the proposed new choice of variables and recalling that we are considering the case of planar waves, we are able to rewrite the governing equations [3.1] as different uncoupled sets of equations, namely:

– a set of three equations only involving longitudinal quantities (left) and two sets of three equations only involving transverse quantities in the k-th direction, with $\xi = 2, 3$ (right):

$$\begin{cases} \ddot{u}_1 = c_p^2 u_{1,11} - \frac{2\mu_e}{\rho} P_{,1}^D - \frac{3\lambda_e + 2\mu_e}{\rho} P_{,1}^S, \\[2mm] \ddot{P}^D = \frac{4}{3}\frac{\mu_e}{\eta} u_{1,1} + \frac{1}{3}c_m^2 P_{,11}^D - \frac{2}{3}c_m^2 P_{,11}^S - \omega_s^2 P^D, \\[2mm] \ddot{P}^S = \frac{3\lambda_e + 2\mu_e}{3\eta} u_{1,1} - \frac{1}{3}c_m^2 P_{,11}^D + \frac{2}{3}c_m^2 P_{,11}^S - \omega_p^2 P^S, \end{cases}$$

$$\begin{cases} \ddot{u}_\xi = c_s^2 u_{\xi,11} - \frac{2\mu_e}{\rho} P_{(1\xi),1} + \frac{\eta}{\rho}\omega_r^2 P_{[1\xi],1}, \\[2mm] \ddot{P}_{(1\xi)} = \frac{\mu_e}{\eta} u_{\xi,1} + \frac{1}{2}c_m^2 P_{(1\xi),11} + \frac{1}{2}c_m^2 P_{[1\xi],11} - \omega_s^2 P_{(1\xi)}, \\[2mm] \ddot{P}_{[1\xi]} = -\frac{1}{2}\omega_r^2 u_{\xi,1} + \frac{1}{2}c_m^2 P_{(1\xi),11} + \frac{1}{2}c_m^2 P_{[1\xi],11} - \omega_r^2 P_{[1\xi]}, \end{cases}$$

[3.8]

– three uncoupled equations only involving the variables $P_{(23)}$, $P_{[23]}$ and P^V, respectively,

---

1 Due to the chosen values of the parameters, which are supposed to satisfy [3.4], all the introduced characteristic velocities and frequencies are real. Indeed, the condition $(3\lambda_e + 2\mu_e) > 0$, together with the condition $\mu_e > 0$, imply the condition $(\lambda_e + 2\mu_e) > 0$.

$$\ddot{P}_{(23)} = -\omega_s^2 P_{(23)} + c_m^2 P_{(23),11}, \quad \ddot{P}_{[23]} = -\omega_r^2 P_{[23]} + c_m^2 P_{[23],11},$$
$$\ddot{P}^V = -\omega_s^2 P^V + c_m^2 P_{,11}^V. \tag{3.9}$$

These 12 scalar differential equations will be used to study wave propagation in our relaxed micromorphic media.

3.1.2. *Plane wave propagation*

We now look for a wave form solution of the previously derived equations of motion. We start from the uncoupled equations [3.9] and assume that the involved unknown variables take the harmonic form:

$$P_{(23)} = Re\left\{\beta_{(23)}e^{i(kX-\omega t)}\right\}, \quad P_{[23]} = Re\left\{\beta_{[23]}e^{i(kX-\omega t)}\right\},$$
$$P^V = Re\left\{\beta^V e^{i(kX-\omega t)}\right\}, \tag{3.10}$$

where $\beta_{(23)}$, $\beta_{[23]}$ and β^V are the amplitudes of the three introduced waves. Replacing this wave form in equations [3.9] and simplifying, we obtain the following dispersion relations, respectively:

$$\omega(k) = \sqrt{\omega_s^2 + k^2 c_m^2}, \quad \omega(k) = \sqrt{\omega_r^2 + k^2 c_m^2}, \quad \omega(k) = \sqrt{\omega_s^2 + k^2 c_m^2}. \tag{3.11}$$

We notice that for a vanishing wave number ($k = 0$), the dispersion relations for the three considered waves give non-vanishing frequencies so that these waves are so-called *optic waves* with cutoff frequencies ω_s, ω_r and ω_s, respectively.

We introduce the unknown vectors $\mathbf{v}_1 = \left(u_1, P^D, P^S\right)$ and $\mathbf{v}_\xi = \left(u_\xi, P_{(1\xi)}, P_{[1\xi]}\right)$, $\xi = 2, 3$ and look for wave form solutions of equations [3.8] in the form:

$$\mathbf{v}_1 = Re\left\{\boldsymbol{\beta}e^{i(kX-\omega t)}\right\}, \quad \mathbf{v}_\xi = Re\left\{\boldsymbol{\gamma}^\xi e^{i(kX-\omega t)}\right\}, \xi = 2, 3, \tag{3.12}$$

where $\boldsymbol{\beta} = (\beta_1, \beta_2, \beta_3)^T$ and $\boldsymbol{\gamma}^\xi = (\gamma_1^\xi, \gamma_2^\xi, \gamma_3^\xi)^T$ are the unknown amplitudes of the considered waves. Replacing this expressions in equations [3.8], we obtain, respectively,

$$\mathbf{A}_1 \cdot \boldsymbol{\beta} = 0, \quad \mathbf{A}_\xi \cdot \boldsymbol{\gamma}^\xi = 0, \quad \xi = 2, 3, \tag{3.13}$$

where

$$
\mathbf{A}_1 =
\begin{pmatrix}
-\omega^2 + c_p^2\, k^2 & i\,k\, 2\mu_e/\rho & i\,k\,(3\lambda_e + 2\mu_e)/\rho \\
-i\,k\,\tfrac{4}{3}\,\mu_e/\eta & -\omega^2 + \tfrac{1}{3}k^2 c_m^2 + \omega_s^2 & -\tfrac{2}{3}\,k^2 c_m^2 \\
-\tfrac{1}{3}\,i\,k\,(3\lambda_e + 2\mu_e)/\eta & -\tfrac{1}{3}\,k^2\,c_m^2 & -\omega^2 + \tfrac{2}{3}\,k^2\,c_m^2 + \omega_p^2
\end{pmatrix},
$$

$$
\mathbf{A}_2 = \mathbf{A}_3 =
\begin{pmatrix}
-\omega^2 + k^2 c_s^2 & i\,k\, 2\mu_e/\rho & -i\,k\,\tfrac{\eta}{\rho}\omega_r^2, \\
-i\,k\, 2\mu_e/\eta, & -2\omega^2 + k^2 c_m^2 + 2\omega_s^2 & k^2 c_m^2 \\
i\,k\,\omega_r^2 & k^2 c_m^2 & -2\omega^2 + k^2 c_m^2 + 2\omega_r^2
\end{pmatrix}.
$$

In order to have non-trivial solutions of the algebraic systems [3.13], we must impose that:

$$
\det \mathbf{A}_1 = 0, \qquad \det \mathbf{A}_2 = 0, \qquad \det \mathbf{A}_3 = 0, \tag{3.14}
$$

which are the so-called dispersion relations $\omega = \omega(k)$ for the longitudinal and transverse waves in the relaxed micromorphic continuum.

3.1.3. *Numerical results*

In this section, following Mindlin [MIN 64, ERI 99], we will show the dispersion relations $\omega = \omega(k)$ associated with our relaxed micromorphic model.

We start by showing in Table 3.1 (left and center), the values of the parameters of the relaxed model used in the performed numerical simulations. In order to make the obtained results better exploitable, we also recall that in [NEF 13, NEF 07], the following homogenized formulas were obtained which relate the parameters of the relaxed model to the macroscopic Lamé parameters λ and μ which are usually measured by means of standard mechanical tests:

$$
\mu_e = \frac{\mu_h\,\mu}{\mu_h - \mu}, \qquad 2\mu_e + 3\lambda_e = \frac{(2\mu_h + 3\lambda_h)\,(2\mu + 3\lambda)}{(2\mu_h + 3\lambda_h) - (2\mu + 3\lambda)}. \tag{3.15}
$$

These relationships imply that the following inequalities are satisfied: $\mu_h > \mu$, $3\lambda_h + 2\mu_h > 3\lambda + 2\mu$. It is clear that, once the values of the parameters of the relaxed models are known, the standard Lamé parameters can be calculated by means of formulas [3.15], which is what was done in Table 3.1 (right). For completeness, we also show in the same table the corresponding Young modulus and Poisson ratio, calculated by means of the standard formulas.

Parameter	Value	Unit
μ_e	200	MPa
$\lambda_e = 2\mu_e$	400	MPa
μ_h	100	MPa
λ_h	100	MPa
$L_c = L_g$	3	mm

Parameter	Value	Unit
$\alpha_c = \mu_e L_c^2$	1.8×10^{-3}	$MPa\,m^2$
$\alpha_g = \mu_e L_g^2$	1.8×10^{-3}	$MPa\,m^2$
ρ'	2500	Kg/m^3
d	2	mm
$\eta = d^2 \rho'$	10^{-2}	Kg/m

Parameter	Value	Unit
λ	82.5	MPa
μ	66.7	MPa
E	170	MPa
ν	0.28	–

Table 3.1. *Values of the parameters of the relaxed model used in the numerical simulations (left and center) and corresponding values of the Lamé parameters and of the Young modulus and Poisson ratio (right)*

It is evident (see equations [3.7]) that, in general, the relative positions of the horizontal asymptotes ω_l and ω_t as well as of the cutoff frequencies ω_s, ω_r and ω_p can vary depending on the values of the constitutive parameters (see equation [3.7]). It can also be checked that, in the case in which $\lambda_e > 0$ and $\lambda_h > 0$, we always have $\omega_p > \omega_s > \omega_t$ and $\omega_l > \omega_t$; we will maintain this hypothesis in the remainder of the book. The relative position of ω_l and of ω_s can vary depending on the values of the parameters λ_h and μ_h. It can be finally recognized that, in order to have a global band gap, the following conditions must be simultaneously satisfied: $\omega_s > \omega_l$ and $\omega_r > \omega_l$. In terms of the constitutive parameters of the relaxed model, we can say that global band gaps can exist, in the case in which we consider positive values for the parameters λ_e and λ_h, if and only if we have, simultaneously:

$$0 < \mu_e < +\infty, \qquad 0 < \lambda_h < 2\mu_e, \qquad \mu_c > \frac{\lambda_h + 2\mu_h}{2} =: \mu_c^0. \qquad [3.16]$$

As far as negative values for λ_e and λ_h are allowed, the conditions for band gaps are not so straightforward as [3.16], but we do not consider this possibility in this note.

Figure 3.1 shows the dispersion relations for the considered relaxed micromorphic continuum for the lower bound $\mu_c = \mu_c^0$.

It can be immediately seen that we cannot identify a frequency band in which overall wave propagation is forbidden. More precisely, for any value of frequency, at least one wave exists (longitudinal, transverse or uncoupled) which propagates inside the considered medium. Things become different as soon as the value of the Cosserat couple modulus increases. In Figure 3.2, we show the dispersion relations for the relaxed micromorphic model corresponding to $\mu_c = 2\,\mu_c^0$. It can be remarked that in this case, any type of wave can propagate in the frequency range $[\omega_l, \omega_s]$, so that we can state that a band gap can be observed which is actually triggered by the increasing value of the Cosserat couple modulus. In this frequency range, the wavenumber becomes imaginary and only standing waves exist. For the sake of completeness, we also show in Figure 3.3, a third value for the Cosserat couple modulus, namely $\mu_c = 3\,\mu_c^0$. It can be easily noticed from Figure 3.3 that the frequency band gap remains unchanged with respect to the previous case, even if the global behavior of the system is much more complicated (two cutoff frequencies for the uncoupled and transverse waves instead of one). Hence, we can confirm that the presence of band gaps is actually triggered by the Cosserat couple modulus and we can observe that the wider extension of such band gaps is reached for $\mu_c = 2\,\mu_c^0$.

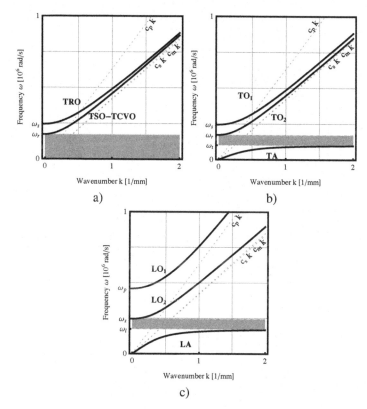

Figure 3.1. *Dispersion relations* $\omega = \omega(k)$ *for* $\mu_c = \mu_c^0$*: uncoupled waves a), longitudinal waves b), transverse waves c)*

We conclude by saying that the relaxed micromorphic model proposed in [GHI 13, NEF 13] is able to describe the presence of frequency band gaps in which no wave propagation can occur. The presence of band gaps is intrinsically related to a critical value of the Cosserat couple modulus μ_c (see [JEO 10, JEO 09, NEF 06, NEF 09, NEF 10] for its interpretation) which must be greater than a threshold value in order to let band gaps appear. This parameter can hence be seen as a discreteness quantifier which starts accounting for lattice discreteness as soon as it reaches the threshold value specified in equation [3.16]. This fact is a novel feature of the introduced relaxed model: we claim that neither the classical micromorphic continuum model nor the Cosserat and the second gradient models are able to predict such band gap phenomena.

We want to finally underline that the fact of disposing of a continuum model which is able to predict the possibility of band gaps in a metamaterial is a very powerful tool for engineering design. Indeed, if we think to tune the few introduced coefficients by means of simple experiments on (already existing) metamaterials exhibiting frequency band gaps, then very complicated engineering structures can be conceived based on such metamaterials which show very peculiar properties with respect to wave propagation and adsorption. In fact, by its own nature, a continuum model is intrinsically versatile for finite element calculations which are, at the current stage of technology, the main tool used by engineers for the design of new structures with enhanced capabilities.

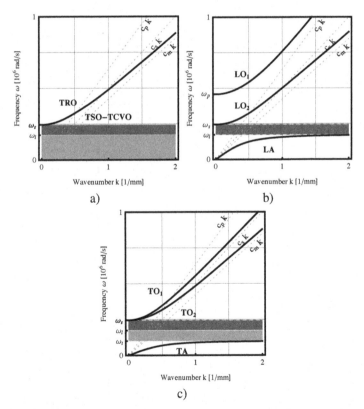

Figure 3.2. *Dispersion relations $\omega = \omega(k)$ for $\mu_c = 2\,\mu_c^0$: uncoupled waves a), longitudinal waves b), transverse waves c)*

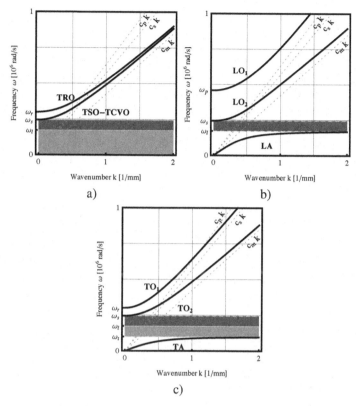

Figure 3.3. *Dispersion relations $\omega = \omega(k)$ for $\mu_c = 3\,\mu_c^0$:*
uncoupled waves a), longitudinal waves b), transverse waves c).

3.2. Reflection and transmission of waves at discontinuity surfaces in second gradient continua

In this section, we present the second aspect concerning wave propagation which we want to address in this book, namely the problem of the study of wave reflection and transmission of harmonic plane waves at discontinuity surfaces between second gradient continua. We recall once again that this problem is conceptually separated from the problem of studying band gaps in relaxed micromorphic continua which has been presented in the previous sections. Actually, in the case of second gradient continua no band gaps can be observed when considering waves propagating in the bulk, since only acoustic waves can be found in such theories and this means that for any frequency, there always exists a real wavenumber associated with a propagative wave. The fact of considering second gradient metamaterials becomes interesting when a discontinuity surface is introduced inside the considered bulk medium and when the phenomena of reflection and transmission are examined. Indeed, playing with three factors, namely (1) the elastic coefficients of the metamaterial, (2) the type of imposed boundary conditions at the interface and (3) the material properties of the interface itself, we can expect to vary the amount of energy which is reflected and transmitted at the interface also arriving to obtain complete reflection and transmission which are the more interesting cases from an engineering point of view. In fact, materials which are able to completely adsorb waves could be used for applications related to noise adsorption or stealth technology. However, metamaterials which are able to completely reflect waves could be used, for example, for acoustic optimization of buildings. Moreover, metamaterials which are known to show microstructure-related peculiar reflection/transmission phenomena could be used for a more effective non-destructive control of damage evolution (see [MAD 14a]).

3.2.1. *Mechanical energy transport in second gradient continua*

In this section we deduce, starting from the appropriate equations of motion, the mechanical energy conservation law for a three-dimensional (3D) second gradient continuum.

Let $\chi : B \times (0, T) \to \mathbb{R}^3$ be the placement map which, at any instant t, associates with any material particle $\mathbf{X} \in B$ its position in the physical space. The displacement field is then defined as $\mathbf{u}(\mathbf{X}, t) := \chi(\mathbf{X}, t) - \mathbf{X}$. We set $\mathbf{F} := \nabla \chi$ and we denote by $\mathbf{E} := (\mathbf{F}^T \cdot \mathbf{F} - \mathbf{I})/2$ the classical Green–Lagrange deformation tensor. Let ρ be the mass per unit volume of the considered continuum in its reference configuration, we introduce, by means of:

$$\mathcal{E} = \frac{1}{2}\rho(\dot{\mathbf{u}})^2 + \Psi(\varepsilon, \nabla \varepsilon),$$ [3.17]

the total Lagrangian energy density of the considered second gradient continuum, as given by the sum of the kinetic and deformation energy, which we denote by Ψ. Here, and in the following, a superposed dot represents partial differentiation with respect to time, i.e. what is usually called the material time derivative. Moreover, we recall that, in absence of body forces, the equation of motion for a second gradient continuum reads[2]:

$$div\left[\mathbf{F} \cdot \left(\frac{\partial \Psi}{\partial \mathbf{E}} - div\left(\frac{\partial \Psi}{\partial \nabla \mathbf{E}}\right)\right)\right] = \rho\ddot{\mathbf{u}}$$ [3.18]

---

2 The symbol div stands for the usual divergence operator, e.g. $(div\mathbf{A})_{ij} = \mathbf{A}_{ijk,k}$. Here, and in the sequel, we adopt Einstein summation convention over repeated indices. The symbol ∇ stands for the usual gradient operator, e.g. $(\nabla\mathbf{A})_{ijk} = \mathbf{A}_{ij,k}$. A central dot indicates a simple contraction between two tensors of any order, e.g. $(\mathbf{A} \cdot \mathbf{B})_{ijhk} = \mathbf{A}_{ijp}\mathbf{B}_{phk}$

Differentiating equation [3.17] with respect to time and using equation [3.18], it can be shown that[3]:

$$\frac{\partial \mathcal{E}}{\partial t} + div\left[-\dot{\mathbf{u}} \cdot \mathbf{F} \cdot \left(\frac{\partial \Psi}{\partial \mathbf{E}} - div\left(\frac{\partial \Psi}{\partial \nabla \mathbf{E}}\right)\right) - \left((\nabla \dot{\mathbf{u}})^T \cdot \mathbf{F}\right) : \frac{\partial \Psi}{\partial \nabla \mathbf{E}}\right] = 0. \quad [3.19]$$

In the calculations for obtaining equation [3.19], we used the fact that ε is a second-order symmetric tensor, that $\nabla \varepsilon$ is a third-order tensor symmetric with respect to its first two indices and that $\nabla \mathbf{F}$ is a third-order tensor symmetric with respect to its last two indices. Thus, this last equation represents the Lagrangian form of energy balance for a second gradient 3D continuum in the general nonlinear case.

We now want to focus our attention on the particular case of linear elasticity in order to study linear plane waves in second gradient 3D continua. To this purpose, we note that, when linearizing in the neighborhood of a stress-free reference configuration, the gradient of placement $\mathbf{F}$ in equation [3.19] is substituted by the identity matrix and that the equation of motion and mechanical energy balance for a second gradient continuum, respectively, reduce to

$$div\left[\mathbf{S} - div\mathbf{\Pi}\right] = \rho\ddot{\mathbf{u}}, \qquad \frac{\partial \mathcal{E}}{\partial t} + div\left[-\dot{\mathbf{u}} \cdot (\mathbf{S} - div\mathbf{\Pi}) - (\nabla \dot{\mathbf{u}})^T : \mathbf{\Pi}\right] = 0, \quad [3.20]$$

where, following the nomenclature of Germain, $\mathbf{S}$ and $\mathbf{\Pi}$ are the linearized Piola–Kirchoff first and second gradient stress tensors, respectively. In order to lighten the notation, however, we will refer to these two tensors simply as stress and hyper-stress tensors, respectively. It is well known that in the case of linear-isotropic materials $\mathbf{S} = 2\mu\varepsilon + \lambda(tr\varepsilon)\mathbf{I}$, where $\varepsilon = (\nabla\mathbf{u} + (\nabla\mathbf{u})^T)/2$ is the linearized Green–Lagrange deformation tensor and λ and μ are the so-called Lamé coefficients. As for the hyper-stress third-order tensor $\mathbf{\Pi}$, it can be shown (see [DEL 09a]) that in the case of isotropic materials it takes the following simplified form[4]:

$$\mathbf{\Pi} = c_2\left[2\mathbf{I} \otimes div\varepsilon + (\mathbf{I} \otimes \nabla(tr\varepsilon))^{T_{23}} + \nabla(tr\varepsilon) \otimes \mathbf{I}\right] + c_3\mathbf{I} \otimes \nabla(tr\varepsilon)$$

$$+ 2c_5\left[(\mathbf{I} \otimes div\varepsilon)^{T_{23}} + div\varepsilon \otimes \mathbf{I}\right] + 2c_{11}\nabla\varepsilon + 4c_{15}(\nabla\varepsilon)^{T_{12}}, \quad [3.21]$$

---

3 A double dot indicates a double contraction between two tensors of any order, e.g. $(\mathbf{A} : \mathbf{B})_{ij} = \mathbf{A}_{ihk}\mathbf{B}_{khj}$.

4 We define the transposition operations of a third-order tensor as $A_{ijk}^{T_{23}} = A_{ikj}$ and $A_{ijk}^{T_{12}} = A_{jik}$ and the symbol $\otimes$ as the usual tensor product operation between two tensors of any order (e.g. $(A \otimes B)_{ijhk} = A_{ij}B_{hk}$)

where c_2, c_3, c_5, c_{11} and c_{15} are constants depending on the material properties of the considered second gradient continuum. As it will be seen to be useful later on, we define the following coefficients

$$\Lambda := c_3 + 2(c_5 + c_{15}) + 4c_2, \qquad M := c_{11} + c_{15} + c_5, \qquad \text{[3.22]}$$

which parallel the first gradient Lamé coefficients λ and μ.

3.2.2. *Dispersion formulas*

Let us now consider a wave traveling in the considered second gradient continuum. We denote by x_1 the axis of a reference frame the direction of which coincides with the propagation direction and by x_2 and x_3 the other two directions forming a Cartesian basis with x_1. We assume that the displacement vector has three non-vanishing components depending only on the x_1 coordinate and on time, i.e. $\mathbf{u}(x_1,t) = (u_1(x_1,t), u_2(x_1,t), u_3(x_1,t))$. In the following, we will say that we are in presence of a unidirectional wave propagation. With this assumption, it is easy to show that the associated matrix form of the linearized deformation tensor reads:

$$\varepsilon = \begin{pmatrix} u_1' & u_2'/2 & u_3'/2 \\ u_2'/2 & 0 & 0 \\ u_3'/2 & 0 & 0 \end{pmatrix} \qquad \text{[3.23]}$$

where we clearly denote by an apex the partial differentiation with respect to the space variable x_1. Using [3.23] to calculate the stress and hyper-stress tensors $\mathbf{S} = 2\mu\varepsilon + \lambda(tr\varepsilon)\mathbf{I}$ and $\mathbf{P}$ (see equations [3.21] and [3.22]), the equation of motion [3.20]$_1$ takes the following form:

$$(\lambda + 2\mu)u_1'' - (\Lambda + 2M)u_1'''' = \rho\ddot{u}_1, \qquad \text{[3.24]}$$

$$\mu u_2'' - M u_2'''' = \rho\ddot{u}_2, \qquad \mu u_3'' - M u_3'''' = \rho\ddot{u}_3. \qquad \text{[3.25]}$$

We notice that the equations of motion obtained in this unidirectional wave propagation are completely uncoupled due to the particular constitutive relations assumed for isotropic, linear elastic, second gradient continua. Moreover, the conservation of energy [3.20]$_2$ gives $\partial\mathcal{E}/\partial t + H' = 0$, where we denote:

$$H := -(\lambda + 2\mu)(u_1'\dot{u}_1) - \mu(u_2'\dot{u}_2) - \mu(u_3'\dot{u}_3) \qquad \text{[3.26]}$$
$$+(\Lambda + 2M)(u_1'''\dot{u}_1 - u_1''\dot{u}_1') + M(u_2'''\dot{u}_2 - u_2''\dot{u}_2') + M(u_3'''\dot{u}_3 - u_3''\dot{u}_3'),$$

the energy flux in the considered particular case. We also remark that M and $\Lambda + 2M$ are positive due to definite-positiveness of the internal energy (see [DEL 09a]). We now assume that the displacement field admits a classical wave solution in the form:

$$\mathbf{u} = \begin{pmatrix} u_1 \\ u_2 \\ u_3 \end{pmatrix} = \begin{pmatrix} \alpha_1 \\ \alpha_2 \\ \alpha_3 \end{pmatrix} e^{i(\omega t - kx_1)} \qquad [3.27]$$

where the eigenvector $(\alpha_1, \alpha_2, \alpha_3)$ gives the longitudinal and transversal amplitudes of the considered wave, ω is the positive real frequency and k is its wave number. Using this wave form for $\mathbf{u}$ in the equations of motion for longitudinal and transversal displacement [3.24] and [3.25], we get the following dispersion relations for a second gradient continuum:

$$(\Lambda + 2M)k_1^4 + (\lambda + 2\mu)k_1^2 - \rho\omega^2 = 0,$$

$$Mk_2^4 + \mu k_2^2 - \rho\omega^2 = 0, \qquad Mk_3^4 + \mu k_3^2 - \rho\omega^2 = 0,$$

where k_1 is the wave number relative to the eigenvector $(1,0,0)$, k_2 and k_3 are the wave numbers relative to the eigenvectors $(0,1,0)$ and $(0,0,1)$, respectively. Since we are dealing with isotropic media, the two transverse dispersion relations coincide; in what follows we will therefore ignore the last dispersion relation.

Because of isotropy, the waves arising in considered medium can be either purely transversal or purely longitudinal. We now look for non-dimensional forms of these relations by setting $k_1 = k_l \tilde{k}_l$, $\omega = \omega_l \tilde{\omega}$ for longitudinal waves and $k_2 = k_t \tilde{k}_t$, $\omega = \omega_t \tilde{\omega}$ for transverse waves. Here k_l (or k_t), ω_l (or ω_t) are characteristic values of the wave number and of the frequency for longitudinal (or transverse) waves, respectively; moreover, $\tilde{k}_l$, $\tilde{k}_t$ and $\tilde{\omega}$ are the corresponding dimensionless variables. This leads to:

$$L_l^2 k_l^2 \tilde{k}_l^4 + \tilde{k}_l^2 - \frac{\rho}{(\lambda + 2\mu)} \frac{\omega_l^2}{k_l^2} \tilde{\omega}^2 = 0,$$

$$L_t^2 k_t^2 \tilde{k}_t^4 + \tilde{k}_t^2 - \frac{\rho}{\mu} \frac{\omega_t^2}{k_t^2} \tilde{\omega}^2 = 0,$$

where $L_l := \sqrt{(\Lambda + 2M)/(\lambda + 2\mu)}$ and $L_t := \sqrt{M/\mu}$ are the characteristic length of second gradient interactions for longitudinal and transversal waves, respectively. To our knowledge, similar dispersion formulas for a particular class of second gradient materials has been already studied only by [TOU 62]. We finally choose k_l^2 and k_t^2 to be such that the coefficients of $\tilde{\omega}^2$ in the two dispersion relations are both equal to one and hence we obtain:

$$k_l^2 = \rho\omega_l^2/(\lambda + 2\mu) \qquad k_t^2 = \rho\omega_t^2/\mu. \qquad [3.28]$$

The dispersion relations thus reduce to:

$$\epsilon_l^2 \tilde{k}_l^4 + \tilde{k}_l^2 - \tilde{\omega}^2 = 0, \qquad \epsilon_t^2 \tilde{k}_t^4 + \tilde{k}_t^2 - \tilde{\omega}^2 = 0, \qquad [3.29]$$

where we set

$$\epsilon_l = k_l\, L_l, \qquad\qquad \epsilon_t = k_t\, L_t.$$

In other words, the introduced quantity ϵ_l (or ϵ_t) is the ratio between the characteristic second gradient length and the wavelength for longitudinal (transverse) waves. The chosen non-dimensional form of the dispersion equations implies that when $\tilde{\omega} = 1$, then the dimensional frequency ω takes the characteristic value $\omega = \omega_l = \epsilon_l L_l \sqrt{(\lambda + 2\mu)/\rho}$ in the case of longitudinal waves and the value $\omega = \omega_t = \epsilon_t L_t \sqrt{\mu/\rho}$ in the case of transverse waves. Analogously, when $\tilde{k} = 1$, then the dimensional wave number takes the characteristic value $k_1 = k_l = \epsilon_l/L_l$ in the case of longitudinal waves and the value $k_2 = k_t = \epsilon_t/L_t$ in the case of transverse waves.

The relationships between the dimensionless wave number and frequency is thus easily recovered both for longitudinal and transverse waves by solving the bi-quadratic equations [3.29]:

$$\tilde{k}_l = \pm \sqrt{\frac{-1 \pm \sqrt{1 + 4\,\epsilon_l^2\,\tilde{\omega}^2}}{2\epsilon_l^2}}, \qquad\qquad \tilde{k}_t = \pm \sqrt{\frac{-1 \pm \sqrt{1 + 4\,\epsilon_t^2\,\tilde{\omega}^2}}{2\epsilon_t^2}}.$$

If we consider the positive real numbers

$$\tilde{k}_l^p = \sqrt{\frac{-1 + \sqrt{1 + 4\,\epsilon_l^2\,\tilde{\omega}^2}}{2\epsilon_l^2}}, \qquad\qquad \tilde{k}_l^s = \sqrt{\frac{1 + \sqrt{1 + 4\,\epsilon_l^2\,\tilde{\omega}^2}}{2\epsilon_l^2}}, \qquad [3.30]$$

$$\tilde{k}_t^p = \sqrt{\frac{-1 + \sqrt{1 + 4\,\epsilon_t^2\,\tilde{\omega}^2}}{2\epsilon_t^2}}, \qquad\qquad \tilde{k}_t^s = \sqrt{\frac{1 + \sqrt{1 + 4\,\epsilon_t^2\,\tilde{\omega}^2}}{2\epsilon_t^2}}, \qquad [3.31]$$

the four roots associated with longitudinal waves are clearly $\pm\tilde{k}_l^p$, $\pm j\tilde{k}_l^s$, and analogously we have for the transverse waves the four roots $\pm\tilde{k}_t^p$, $\pm j\tilde{k}_t^s$, where j stands for the imaginary unit. It is clear that we can derive the corresponding dimensional quantities k_1^p and k_1^s (for longitudinal waves) and k_2^p and k_2^s (for transverse waves) just by multiplying the non-dimensional roots $\tilde{k}_l$ and $\tilde{k}_t$ by k_l and k_t, respectively (see equation [3.28]). As we will see in more detail later on, the roots $\tilde{k}^p$ are associated with *propagative waves* which are a second gradient generalization of the waves propagating in a first gradient material, while the roots $\tilde{k}^s$ are the so-called *standing waves* (or evanescent waves) and are peculiar of second gradient models. These standing waves will be seen to play a significant role close to material discontinuity surfaces in second gradient continua where phenomena of reflection and transmission may occur. Indeed, there are other physical situations in which standing waves may appear. For instance, [OUI 03] showed that this may occur when studying coupling between the transversal displacement u and the longitudinal displacement w in 1D beams.

We explicitly remark that the dispersion relations [3.29] are very different from the relations obtained in the case of relaxed micromorphic continua and discussed in the previous sections. In fact, if we try to plot the roots of such relations corresponding to a positive direction of propagation, then the result qualitatively presented in Figure 3.4 is obtained both for longitudinal and transverse waves. One propagative wave and one standing wave can be recognized for any direction of propagation.

This means that for any real frequency $\tilde{\omega}$, there always exists a real wavenumber $\tilde{k}_l$ (or $\tilde{k}_t$) which allows for global wave propagation inside the material. As a consequence, no band gaps can be observed in the bulk when considering wave propagation inside a second gradient material. This result is also obtained in [MAD 13] where a second gradient medium is regarded as a suitable limit case of the previously introduced relaxed micromorphic medium.

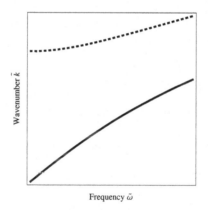

Figure 3.4. *Dispersion relation for a second gradient continuum: acoustic (longitudinal or transverse) wave*

We will see in the following that, despite the fact that no band gaps are possible in second gradient media, reflection and transmission at discontinuity surfaces are heavily influenced from the presence of microstructure especially at high frequencies.

3.2.3. *Natural and kinematical boundary conditions at surfaces where displacement or normal derivative of displacement may be discontinuous*

The problem of finding boundary conditions to be imposed at discontinuity surfaces is always challenging for the modeling of any physical phenomenon. As a matter of fact, the only possible method which leads to surely well-posed boundary conditions, compatible with the obtained bulk equations of motion, is to use a variational principle. This is also the case when looking for the correct set of boundary conditions associated with the motion of a second gradient continuum. It can be checked, following the variational arguments presented, e.g. in [DEL 09a, SCI 08], that the boundary conditions on the considered discontinuity surface S, in absence of external surface and line actions, can be deduced from the following duality conditions

$$[|\mathbf{t} \cdot \delta \mathbf{u}|] = 0, \qquad [|\boldsymbol{\tau} \cdot (\delta \mathbf{u})_n|] = 0, \qquad [|\mathbf{f} \cdot \delta \mathbf{u}|] = 0. \qquad [3.32]$$

The first two of these conditions are valid on S while the last one is valid on the edges of S, if any. In the previous formulas [3.32], we set

$$\mathbf{t} := \left[\mathbf{F} \cdot \left(\frac{\partial \Psi}{\partial \mathbf{E}} - div \left(\frac{\partial \Psi}{\partial \nabla \mathbf{E}} \right) \right) \right] \cdot \mathbf{n} - div^S \left(\mathbf{F} \cdot \frac{\partial \Psi}{\partial \nabla \mathbf{E}} \cdot \mathbf{n} \right),$$

$$\boldsymbol{\tau} := \left(\mathbf{F} \cdot \frac{\partial \Psi}{\partial \nabla \mathbf{E}} \cdot \mathbf{n} \right) \cdot \mathbf{n}, \qquad \mathbf{f} := \left(\mathbf{F} \cdot \frac{\partial \Psi}{\partial \nabla \mathbf{E}} \cdot \mathbf{n} \right) \cdot \boldsymbol{\nu}.$$

Moreover, $\mathbf{n}$ is the unit normal vector to the surface S, div^S stands for the surface divergence operator on S, if the edge is regarded as the border of a surface then $\boldsymbol{\nu}$ is the normal vector to the considered edge which is tangent to the surface, $\delta \mathbf{u}$ is the variation of the displacement field and $(\delta \mathbf{u})_n := \nabla(\delta \mathbf{u}) \cdot \mathbf{n}$ stands for the normal derivative of the variation of the displacement field. Finally, given a quantity a defined everywhere and having continuous traces a^+ and a^- on the two sides of S respectively, we have set $[|a|] := a^+ - a^-$ (we use the same symbol for the jump across edges).

When choosing arbitrary displacement variation $\delta \mathbf{u}$ continuously varying through S and arbitrary normal derivative $(\delta \mathbf{u})_n$ continuously varying through S in equations [3.32], we get a particular set of *natural jump conditions*, which can be interpreted as the vanishing jump of internal surface forces ($[|\mathbf{t}|] = 0$), the vanishing jump of internal surface *double forces* ($[|\boldsymbol{\tau}|] = 0$) and the vanishing jump of internal line actions ($[|\mathbf{f}|] = 0$), respectively (see [GER 73a, GER 73b], for the first introduction of the concept of contact double force). Actually, if it happens that the displacement field, due to a particular constraint, verifies the particular equation $\delta \mathbf{u}^- = \delta \mathbf{u}^+ =: \delta \mathbf{u}$ (vanishing jump of the displacement field), and $\delta \mathbf{u}$ is arbitrary, then in order to fulfill conditions $[3.32]_1$ and $[3.32]_3$ we must require that also the jumps of the dual quantities to $\delta \mathbf{u}$ (i.e. surface forces and line forces) are vanishing. Analogously, if the normal derivative of displacement is assigned to be equal on both sides of S and this common value is arbitrary, then in order equation $[3.32]_2$ to be verified, we must also impose continuity of internal double forces. Conditions of the type $\delta \mathbf{u}^- = \delta \mathbf{u}^+$, or any similar relationship among $\delta \mathbf{u}^\pm$ and $(\delta \mathbf{u})_n^\pm$, are called *kinematical boundary conditions.* Once the kinematical boundary conditions are chosen, the associated dual conditions, necessary and sufficient to fulfill duality conditions [3.32], are denoted natural boundary conditions associated with the chosen kinematical conditions. Therefore, in addition to those previously discussed, other kinematical choices are possible in the previously formulated duality conditions, these choices being indicated by the admissible kinematics of the considered system or, in simpler words, by the considered phenomenology. We discuss here some kinematical constraints for second gradient continua, which we call *generalized internal clamp, generalized internal hinge* and *generalized internal roller* and the natural boundary conditions associated with them:

– Generalized internal clamp

We define this constraint imposing the continuity of both the displacement $\mathbf{u}$ and the normal derivative of displacement $\nabla \mathbf{u} \cdot \mathbf{n}$ (and therefore of the test function $\delta \mathbf{u}$ and of its normal derivative $(\delta \mathbf{u})_n$) at discontinuity surface S which from now on we assume to be regular (i.e S has no edges). As already noticed, in this case, the boundary conditions to be imposed are:

$$[|\mathbf{u}|] = 0, \qquad [|\nabla \mathbf{u} \cdot \mathbf{n}|] = 0, \qquad [|\mathbf{t}|] = 0, \qquad [|\boldsymbol{\tau}|] = 0. \qquad [3.33]$$

If we consider unidirectional wave propagation, and if we choose a flat surface S such that its normal is given by $\mathbf{n} = (1, 0, 0)$, then the jump conditions [3.33] particularize into

$$[|u_1|] = 0, \quad [|u_1'|] = 0, \quad [|(\lambda + 2\mu)u_1' - (\Lambda + 2M)u_1''')|] = 0,$$
$$[|(\Lambda + 2M)u_1''|] = 0,$$
$$[|u_2|] = 0, \quad [|u_2'|] = 0, \quad [|(\mu u_2' - Mu_2''')|] - 0, \quad [|Mu_2''|] = 0, \qquad [3.34]$$
$$[|u_3|] = 0, \quad [|u_3'|] = 0, \quad [|(\mu u_3' - Mu_3''')|] = 0, \quad [|Mu_3''|] = 0.$$

– Generalized internal elastic hinge

We introduce the constraint of generalized internal elastic hinge at surface S assuming that there exists a surface deformation energy density Ψ_S localized on S, quadratically dependent on the surface relative displacement. In other words, we assume that $\mathbf{u}^+$ and $\mathbf{u}^-$ can be different, and can vary independently. In formulas, the introduced surface deformation energy density is given by:

$$\Psi_S = \frac{1}{2}k_S^n(\mathbf{u}^+ \cdot \mathbf{n} - \mathbf{u}^- \cdot \mathbf{n})^2 + \frac{1}{2}k_S^{\parallel}(u_{\parallel}^+ - u_{\parallel}^-)^2, \qquad [3.35]$$

where we denoted by $u_{\parallel}$ the tangential component of the displacement $\mathbf{u}$. When both surface elastic moduli k_S^n and $k_S^{\parallel}$ tend to infinity, this energy will impose continuity of displacements at S. We will assume that $(\delta \mathbf{u})_n^+$ and $(\delta \mathbf{u})_n^-$ can independently take arbitrary values on the two sides of S. These last conditions, together with the duality condition [3.31]$_2$, imply that the double forces must be separately vanishing on the two sides of S. In formulas, the four conditions for a generalized elastic hinge can be seen to read:

$$[|\mathbf{t}|] = 0, \quad k_S^n \, [|\mathbf{u} \cdot \mathbf{n}|] = \mathbf{t}^+ \cdot \mathbf{n}, \quad k_S^{\parallel} \, [|u_{\parallel}|] = t_{\parallel}^+, \quad \tau^+ = 0, \quad \tau^- = 0, \quad [3.36]$$

where we denoted by $t_{\parallel}^+$ the tangential component of the force $\mathbf{t}^+$ on the $+$ side of S. More complex surface contact phenomena may be eventually modeled by means of more sophisticated boundary conditions. In the considered unidirectional wave propagation, these conditions reduce to:

$$[|(\lambda + 2\mu)u_1' - (\Lambda + 2M)u_1''')|] = 0,$$
$$k_S^n \, [|u_1|] = ((\lambda + 2\mu)u_1' - (\Lambda + 2M)u_1''')^+, \qquad\qquad [3.37]$$
$$((\Lambda + 2M)u_1'')^+ = 0, \qquad ((\Lambda + 2M)u_1'')^- = 0$$
$$[|(\mu u_2' - M u_2''')|] = 0, \qquad k_S^{\parallel} \, [|u_2|] = (\mu u_2' - M u_2''')^+,$$
$$(M u_2'')^+ = 0, \qquad (M u_2'')^- = 0,$$
$$[|(\mu u_3' - M u_3''')|] = 0, \qquad k_S^{\parallel} \, [|u_3|] = \mu u_3^{+\prime} - M u_3^{+\prime\prime\prime}$$
$$(M u_3'')^+ = 0, \qquad (M u_3'')^- = 0.$$

– *Generalized 3D internal roller*

We define the last kind of kinematical constraint considered in this manuscript which we call generalized roller and which is defined in such a way that it allows independently arbitrary displacements on both sides of S (which implies separately $\delta \mathbf{u}^+$ and $\delta \mathbf{u}^-$ to be arbitrary) and such that the normal derivative of displacement is continuous through S (i.e. $[|\nabla \mathbf{u} \cdot \mathbf{n}|] = 0$). The first condition together with equation [3.32]$_1$ implies that the generalized force must be separately vanishing on both sides of S, while the second condition implies continuity of double forces through S. Hence the four conditions for a generalized roller are:

$$\mathbf{t}^+ = 0, \qquad \mathbf{t}^- = 0, \qquad [|\nabla \mathbf{u} \cdot \mathbf{n}|] = 0, \qquad [|\tau|] = 0. \qquad [3.38]$$

It can be shown that in the considered unidirectional propagation case these equations simplify into:

$$((\lambda + 2\mu)u_1' - (\Lambda + 2M)u_1''')^{\pm} = 0, \quad [|u_1'|] = 0, \quad [|(\Lambda + 2M)u_1''|] = 0,$$
$$(\mu u_2' - M u_2''')^+ = (\mu u_2' - M u_2''')^- = 0, \quad [|u_2'|] = 0, \quad [|M u_2''|] = 0,$$
$$\qquad\qquad\qquad\qquad\qquad\qquad\qquad\qquad\qquad\qquad\qquad [3.39]$$
$$(\mu u_3' - M u_3''')^+ = (\mu u_3' - M u_3''')^- = 0, \quad [|u_3'|] = 0, \quad [|M u_3''|] = 0.$$

As occurs for the dispersion relations, also for the boundary conditions, we can note that in the considered linear-elastic case, they are completely uncoupled in the longitudinal and transversal displacement due to isotropy and to the fact that the displacement only depends on the x_1 variable.

3.2.4. Transmission and reflection at discontinuity surfaces

Let us now recall that we are considering a flat discontinuity surface S inside the considered second gradient continuum. We denote by n the unit normal to S and we choose the fixed reference frame in such a way that n points in the x_1 direction: we are assuming that the vector n is always the same at any point of S (see Figure 3.5). As before, we denote by x_2 and x_3 the other two directions of the fixed reference frame forming a Cartesian basis with x_1.

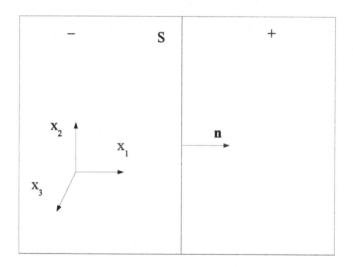

Figure 3.5. *Domain with flat discontinuity surface*

Using the equations of motion [3.24] and [3.25] on both sides of S and the linearized jump conditions [3.34], [3.37] or [3.39], we are able to describe the motion of two different isotropic, linear-elastic second gradient continua which are in contact through the discontinuity surface S and considering three different constraints at this surface (generalized internal clamp, generalized internal hinge and generalized internal roller). Nevertheless, in this manuscript we limit ourselves to the case of two second gradient continua with the same material properties (same first and second gradient elasticity parameters) in contact through the surface S at which we impose the three generalized types of constraints discussed before.

Let us start by studying the case of longitudinal waves (the involved field is then the field we previously denoted by u_1) impacting at the interface S and then we will repeat the reasoning for transverse waves (the involved field is then u_2): this is attributed to the fact that the obtained problem is completely uncoupled with respect to longitudinal and transversal displacements.

Then, let us consider an incident longitudinal wave u_i^l propagating in the x_1 direction and defined as:

$$u_i^l = \alpha_i^l e^{j(\omega t - k_1^P x_1)},$$

where α_i^l is the amplitude of the incident (subscript i), longitudinal (superscript l) wave which we assume to be assigned, ω is the real positive frequency of such an incident wave and, $k_1^P = \tilde{k}_l^P k_l$ (see equations [3.30]$_1$ and [3.28]) is the positive real wave number associated with the propagating wave traveling in the x_1 direction. When this wave reaches the interface S, reflection and transmission phenomena take place (see Figure 3.6).

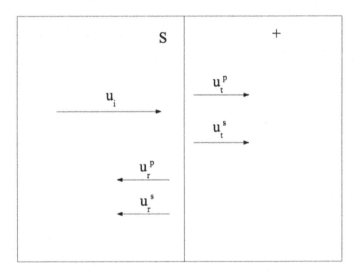

Figure 3.6. *Reflected and transmitted waves at a discontinuity surface between two second gradient materials*

We denote u_r^l and u_t^l the reflected and transmitted longitudinal waves, respectively. According to the geometry of the considered problem and to equation

[3.30], the reflected wave propagates always in the x_1 direction and take the following form:

$$u_r^l = u_r^{lp} + u_r^{ls} = \alpha_r^l e^{j(\omega t + k_1^p x_1)} + \beta_r^l e^{j(\omega t - jk_1^s x_1)},$$

where α_r^l is the amplitude of the propagative reflected wave traveling in the $-x_1$ direction and β_r^l is the amplitude of the standing reflected wave. Note that $k_1^s = \tilde{k}_l^s k_l$ (see equations [3.30]$_2$ and [3.28]) is the positive real wave number associated with the standing wave and that the standing reflected wave vanishes as x_1 approaches $-\infty$. Analogously, the transmitted wave is may be represented by:

$$u_t^l = u_t^{lp} + u_t^{ls} = \alpha_t^l e^{j(\omega t - k_1^p x_1)} + \beta_t^l e^{j(\omega t + jk_1^s x_1)},$$

where α_t^l is the amplitude of the propagative transmitted wave traveling in the x_1 direction and β_t^l is the amplitude of the standing transmitted wave. Note that standing transmitted wave vanishes as x_1 approaches $+\infty$. We remark that, among the two possible standing reflected (transmitted) waves, we did not consider that wave which diverges as $x_1 \to -\infty$ ($x_1 \to +\infty$), as we assume a Sommerfeld-type condition at infinity.

As we want to deal with dimensionless quantities, we introduce the non-dimensional counterpart of these displacements by considering the non-dimensional variables $\tilde{t} = \omega_l t$ and $\tilde{x}_1 = k_l x_1$, where the characteristic quantities ω_l and k_l have been defined in section 3.2.2. We can then introduce the non-dimensional form of the considered displacements as:

$$\tilde{u}_i^l := \frac{u_i^l}{\alpha_i^l} = e^{j(\tilde{\omega}\tilde{t} - \tilde{k}_l^p \tilde{x}_1)}, \tag{3.40}$$

$$\tilde{u}_r^l := \frac{u_r^l}{\alpha_i^l} = \tilde{\alpha}_r^l e^{j(\tilde{\omega}\tilde{t} + \tilde{k}_l^p \tilde{x}_1)} + \tilde{\beta}_r^l e^{j(\tilde{\omega}\tilde{t} - j\tilde{k}_l^s \tilde{x}_1)}, \tag{3.41}$$

$$\tilde{u}_t^l := \frac{u_t^l}{\alpha_i^l} = \tilde{\alpha}_t^l e^{j(\tilde{\omega}\tilde{t} - \tilde{k}_l^p \tilde{x}_1)} + \tilde{\beta}_t^l e^{j(\tilde{\omega}t + j\tilde{k}_l^s \tilde{x}_1)}, \tag{3.42}$$

where clearly we set $\tilde{\alpha}_r^l = \alpha_r^l/\alpha_i^l$, $\tilde{\beta}_r^l = \beta_r^l/\alpha_i^l$, $\tilde{\alpha}_t^l = \alpha_t^l/\alpha_i^l$ and $\tilde{\beta}_t^l = \beta_t^l/\alpha_i^l$, where $\tilde{k}_l^p$ and $\tilde{k}_l^s$ are given by equation [3.30] and where the dimensionless frequency $\tilde{\omega} = \omega/\omega_l$ has been defined in section 3.2.2.

It is clear that once the frequency and the first and second gradient elasticity parameters are given, the only unknowns of the reflection/transmission problem are the four amplitudes $\tilde{\alpha}_r^l$, $\tilde{\beta}_r^l$, $\tilde{\alpha}_t^l$ and $\tilde{\beta}_t^l$. These four scalar unknowns can be found for each of the three types of constraints introduced by considering equations $[3.34]_1$, $[3.37]_1$ and $[3.39]_1$, respectively, which only involve the longitudinal displacement. In conclusion, if we notice that here the first component of $[|u|]$ takes the form $[|u_1|] = u_i^l - u_r^l - u_t^l$, and if we replace the wave form of the longitudinal displacement field in the non-dimensional version of equations $[3.34]_1$, $[3.37]_1$ and $[3.39]_1$ at $x_1 = 0$, we can finally recover the expression of the four amplitudes for each of the three types of constraints considered.

– *Generalized internal clamp*

As for the generalized internal clamp, the non-dimensional jump conditions [3.34] in the considered case in which the two continua on the two sides of S have the same material properties simply read:

$$[|\tilde{u}_1|] = 0, \quad [|\tilde{u}_1'|] = 0, \quad [|\tilde{u}_1' - \epsilon_l^2 \tilde{u}_1'''|] = 0, \quad [|\tilde{u}_1''|] = 0, \qquad [3.43]$$

where, with a slight abuse of notation, we indicate with an apex the derivation operation with respect to the dimensionless space variable $\tilde{x}_1$. Replacing the aforementioned wave form for non-dimensional displacements in these boundary conditions, we calculate the following non-dimensional amplitudes for longitudinal waves:

$$\tilde{\alpha}_r^l = 0, \quad \tilde{\beta}_r^l = 0, \quad \tilde{\alpha}_t^l = 1, \quad \tilde{\beta}_t^l = 0. \qquad [3.44]$$

This means that in the case of a perfect internal clamp at the surface S, the incident wave is completely transmitted. This result is not astonishing if we think that the two materials on both sides of S have been chosen to have the same material properties: indeed, it is as if there were no discontinuity at all and hence the incident wave proceeds unperturbed across the surface S.

– *Generalized elastic internal hinge.*

As for the generalized elastic internal hinge between two continua with the same material properties, the non-dimensional form in the normal direction of jump conditions [3.37] reads:

$$\tilde{k}_S^n [|\tilde{u}_1|] = (\tilde{u}_1' - \epsilon_l^2 u_1''')^+, \quad [|\tilde{u}_1' - \epsilon_l^2 u_1'''|] = 0,$$
$$(\tilde{u}_1'')^+ = 0, \quad (\tilde{u}_1'')^- = 0, \qquad [3.45]$$

where we introduced the dimensionless rigidity $\tilde{k}_S^n = k_S^n / (k_l(\lambda + 2\mu))$. This non-dimensional choice implies that when $\tilde{k}_S^n = 1$, then the dimensional rigidity takes the value $k_S^n = k_l(\lambda + 2\mu)$. The physical meaning of the rigidity $\tilde{k}_S^n$ can be intuitively obtained by thinking to the considered constraint as a series of elastic springs (of rigidity $\tilde{k}_S^n$) joining the two sides of the discontinuity. When $\tilde{k}_S^n$ goes to infinity, then the springs become infinitely stiff so that the displacement on the two sides of the discontinuity takes the same value and the introduced constraint particularizes to a generalized internal hinge. We explicitly remark that the non-dimensional form of the rigidity previously introduced is just one possible choice: this choice does not affect the physical interpretation of the limiting value of $\tilde{k}_S^n$ which we have just given.

It is easy to prove that in the limit case of the elastic constant $\tilde{k}_S^n$ to infinity these boundary conditions reduce to:

$$[[\tilde{u}_1]] = 0, \quad [[\tilde{u}_1' - \epsilon_l^2 u_1''']] = 0, \quad (\tilde{u}_1'')^+ = 0, \quad (\tilde{u}_1'')^- = 0. \qquad [3.46]$$

We will refer to this limit case as *generalized internal hinge* in virtue of the continuity of displacement which is imposed at the discontinuity surface. This constraint differs from the *generalized internal clamp* defined by means of equations [3.43] because the double force (second derivative of displacement) is zero on both sides of the discontinuity, while it can take an arbitrary value in the case of the clamp (even if this value must be the same on both sides). The similarities of the considered 3D constraints with the classical internal hinge and clamp for Euler–Bernoulli beams can be immediately pointed out. As far as the generalized internal hinge is considered, equation [3.46]$_1$ (continuity of displacement) implies by duality the continuity of force (equation [3.46]$_2$), while the double forces (which parallels couples in classical beam theory) are arbitrary on both sides (equations [3.46]$_3$ and [3.46]$_4$). As for the generalized internal clamp, it is easy to recognize that equations [3.43]$_1$ and [3.43]$_3$ parallel the continuity of displacement and force, respectively, and that equation [3.43]$_4$ gives continuity of double forces (continuity of couples in the case of Euler–Bernoulli beams) which also implies equation [3.43]$_2$ by duality.

Using the considered wave form for displacements in boundary conditions [3.45], the calculated non-dimensional amplitudes take the form:

$$\tilde{\alpha}_r^l = \frac{\tilde{k}_l^p(\tilde{k}_l^p - j\tilde{k}_l^s)\{\tilde{k}_l^s\left[\tilde{k}_l^s\epsilon_l^2\left(\tilde{k}_l^o(\epsilon_l\tilde{k}_l^p)^2 + 2\tilde{k}_S^n\right) + 1\right] - 2\tilde{k}_S^n\}}{(\tilde{k}_l^p + j\tilde{k}_l^s)(\tilde{k}_l^p\tilde{k}_l^s\epsilon_l^2 - j)\left[(\tilde{k}_l^p\epsilon_l\tilde{k}_l^s)^2 + 2\tilde{k}_S^n\tilde{k}_l^s - j\tilde{k}_l^p(\tilde{k}_l^s - 2\tilde{k}_S^n)\right]}$$

$$\tilde{\beta}_r^l = \frac{2(\tilde{k}_l^p)^2\left(\epsilon_l^2(\tilde{k}_l^p)^2 + 1\right)\left[(\tilde{k}_l^p\epsilon_l\tilde{k}_l^s)^2 + \tilde{k}_S^n\tilde{k}_l^s - j\tilde{k}_l^p(\tilde{k}_l^s - \tilde{k}_S^n)\right]}{(\tilde{k}_l^p + j\tilde{k}_l^s)\tilde{k}_l^s(\tilde{k}_l^p\tilde{k}_l^s\epsilon_l^2 - j)\left[(\tilde{k}_l^p\epsilon_l\tilde{k}_l^s)^2 + 2\tilde{k}_S^n\tilde{k}_l^s - j\tilde{k}_l^p(\tilde{k}_l^s - 2\tilde{k}_S^n)\right]}$$

$$\tilde{\alpha}_t^l = \frac{2\tilde{k}_l^s(\tilde{k}_l^s + j\tilde{k}_l^p)\tilde{k}_S^n\left[(\epsilon_l\tilde{k}_l^p)^2 + 1\right]}{(\tilde{k}_l^p + j\tilde{k}_l^s)(\tilde{k}_l^p\tilde{k}_l^s\epsilon_l^2 - j)\left[(\tilde{k}_l^p\epsilon_l^2\tilde{k}_l^s)^2 + 2\tilde{k}_S^n\tilde{k}_l^s - j\tilde{k}_l^p(\tilde{k}_l^s - 2\tilde{k}_S^n)\right]}$$

$$\tilde{\beta}_t^l = \frac{2(\tilde{k}_l^p)^2(\tilde{k}_l^s + j\tilde{k}_l^p)\tilde{k}_S^n\left((\epsilon_l\tilde{k}_l^p)^2 + 1\right)}{(\tilde{k}_l^p + j\tilde{k}_l^s)\tilde{k}_l^s(\tilde{k}_l^p\tilde{k}_l^s\epsilon_l^2 - j)\left[(\tilde{k}_l^p\epsilon_l\tilde{k}_l^s)^2 + 2\tilde{k}_S^n\tilde{k}_l^s - j\tilde{k}_l^p(\tilde{k}_l^s - 2\tilde{k}_S^n)\right]}$$

In the limit of elastic constant $\tilde{k}_S^n$ to infinity we have the case of the internal hinge and the calculated non-dimensional amplitudes take the following simplified form:

$$
\tilde{\alpha}_r^l = \frac{j\tilde{k}_l^p[1 - \epsilon_l^2(\tilde{k}_l^s)^2]}{(\tilde{k}_l^p + j\tilde{k}_l^s)(-j + \epsilon_l^2\tilde{k}_l^p\tilde{k}_l^s)}, \quad
\tilde{\beta}_r^l = \frac{(\tilde{k}_l^p)^2[1 + \epsilon_l^2(\tilde{k}_l^p)^2]}{\tilde{k}_l^s(\tilde{k}_l^p + j\tilde{k}_l^s)(-j + \epsilon_l^2\tilde{k}_l^p\tilde{k}_l^s)},
$$

$$
\tilde{\alpha}_t^l = \frac{\tilde{k}_l^s[1 + \epsilon_l^2(\tilde{k}_l^p)^2]}{(\tilde{k}_l^p + j\tilde{k}_l^s)(-j + \epsilon_l^2\tilde{k}_l^p\tilde{k}_l^s)}, \quad
\tilde{\beta}_t^l = \frac{(\tilde{k}_l^p)^2[1 + \epsilon_l^2(\tilde{k}_l^p)^2]}{\tilde{k}_l^s(\tilde{k}_l^p + j\tilde{k}_l^s)(-j + \epsilon_l^2\tilde{k}_l^p\tilde{k}_l^s)}.
$$

$$[3.47]$$

– Generalized 3D internal roller

Finally, if we consider a generalized roller between two continua with the same material properties, the non-dimensional form of jump conditions [3.39] reads

$$
[|\tilde{u}_1'|] = 0, \quad (\tilde{u}_1' - \epsilon_l^2 u_1''')^+ = 0, \quad (\tilde{u}_1' - \epsilon_l^2 u_1''')^- = 0, \quad [|\tilde{u}_1''|] = 0,
$$

Using the considered wave form for dimensionless displacements in these boundary conditions, we get the following values for non-dimensional amplitudes

$$
\tilde{\alpha}_r^l = \frac{k_l^s[1 + \epsilon_l^2(\tilde{k}_l^p)^2]}{(\tilde{k}_l^p + j\tilde{k}_l^s)(-j + \epsilon_l^2\tilde{k}_l^p\tilde{k}_l^s)}, \quad
\tilde{\beta}_r^l = \frac{(\tilde{k}_l^p)^2[1 + \epsilon_l^2(\tilde{k}_l^p)^2]}{\tilde{k}_l^s(\tilde{k}_l^p + j\tilde{k}_l^s)(-j + \epsilon_l^2\tilde{k}_l^p\tilde{k}_l^s)},
$$

$$
\tilde{\alpha}_t^l = \frac{j\tilde{k}_l^p[-1 + \epsilon_l^2(\tilde{k}_l^s)^2]}{(\tilde{k}_l^p + j\tilde{k}_l^s)(-j + \epsilon_l^2\tilde{k}_l^p\tilde{k}_l^s)}, \quad
\tilde{\beta}_t^l = \frac{-(\tilde{k}_l^p)^2[1 + \epsilon_l^2(\tilde{k}_l^p)^2]}{\tilde{k}_l^s(\tilde{k}_l^p + j\tilde{k}_l^s)(-j + \epsilon_l^2\tilde{k}_l^p\tilde{k}_l^s)}.
$$

$$[3.48]$$

Let us now consider the case of transverse waves traveling in the x_1 direction. As done for longitudinal waves, we introduce the incident transversal displacement as:

$$
u_i^t = \alpha_i^t e^{j(\omega t - k_2^p x_1)},
$$

where α_i^t is the amplitude of the incident transverse wave which we assume to be known and k_2^p is the wave number associated with a transverse wave propagating in the x_1 direction. We then call u_r^t and u_t^t the reflected and transmitted transverse wave, respectively. According to the geometry of the considered problem, the reflected and transmitted waves take the following form:

$$
u_r^t = u_r^{tp} + u_r^{ts} = \alpha_r^t e^{j(\omega t + k_2^p x_1)} + \beta_r^t e^{j(\omega t + jk_2^s x_1)},
$$

where α_r^t is the amplitude of the propagative reflected wave traveling in the $-x_1$ direction and β_r^t is the amplitude of the standing reflected wave. Note that the standing reflected wave goes to zero as x_1 approaches $-\infty$. Moreover, the transmitted wave is defined as:

$$u_t^t = u_t^{tp} + u_t^{ts} = \alpha_t^t e^{j(\omega t - k_2^p x_1)} + \beta_t^t e^{j(\omega t - j k_2^s x_1)},$$

The obtained equations for transversal waves are formally analogous to longitudinal waves, since we have just to replace $\tilde{u}_1$ with $\tilde{u}_2$ and all the subscripts and superscripts l with the subscripts and superscripts t, respectively. Repeating the calculations performed for longitudinal waves, we can then calculate the four amplitudes $\tilde{\alpha}_r^t$, $\tilde{\beta}_r^t$, $\tilde{\alpha}_t^t$ and $\tilde{\beta}_t^t$ of the reflected and transmitted transverse waves for the three considered types of constraints imposed at the surface S.

3.2.5. Dependence of transmission and reflection coefficients on second gradient elastic moduli

In this section, we will show plots displaying the reflection and transmission coefficients at discontinuity surfaces of the three types considered before as functions of second gradient elastic moduli. To do so, we start by noticing that owing to the fact that the problem is completely uncoupled in the longitudinal and transversal displacements, we can account separately for the energy fluxes of longitudinal and transversal waves. Moreover, due to the linearity of the problem in study, we can consider separate contributions for the fluxes of the incident, reflected and transmitted waves which, starting from expression [3.26], can be written in terms of the introduced non-dimensional variables, respectively, as:

$$\tilde{H}_i^l(\tilde{x}_1, \tilde{t}) = (\alpha_i^l)^2 k_l \omega_l (\lambda + 2\mu) \left[\ \left(\tilde{u}_i^l \right)' \ \dot{\tilde{u}}_i^l + \epsilon_l^2 \left(\left(\tilde{u}_i^l \right)''' \ \dot{\tilde{u}}_i^l - \left(u_i^l \right)'' \left(\dot{\tilde{u}}_i^l \right)' \right) \right],$$

$$\tilde{H}_r^l(\tilde{x}_1, \tilde{t}) = (\alpha_i^l)^2 k_l \omega_l (\lambda + 2\mu) \left[-\left(\tilde{u}_r^l \right)' \ \dot{\tilde{u}}_r^l + \epsilon_l^2 \left(\left(\tilde{u}_r^l \right)''' \ \dot{\tilde{u}}_r^l - \left(\tilde{u}_r^l \right)'' \left(\dot{\tilde{u}}_r^l \right)' \right) \right],$$

$$\tilde{H}_t^l(\tilde{x}_1, \tilde{t}) = (\alpha_i^l)^2 k_l \omega_l (\lambda + 2\mu) \left[-\left(\tilde{u}_t^l \right)' \ \dot{\tilde{u}}_t^l + \epsilon_l^2 \left(\left(\tilde{u}_t^l \right)''' \ \dot{\tilde{u}}_t^l - \left(\tilde{u}_t^l \right)'' \left(\dot{\tilde{u}}_t^l \right)' \right) \right],$$

$$[3.49]$$

where we recall that, with a slight abuse of notation, the apex and the dot stand here for the derivation operations with respect to the introduced dimensionless space and time variables. A simple inspection of equations [3.49], recalling that the displacement field take the wave form [3.27], shows that for higher frequencies and shorter wavelengths (i.e. larger wave number k) the amount of energy transported because of second gradient effects increases. The corresponding energy fluxes for transversal waves are formally analogous and are obtained from the previous ones just replacing everywhere the apex l with the apex t and the parameter ϵ_l with the parameter ϵ_t. In order to be able to calculate the reflection and transmission coefficients for the considered problem, we substitute in equations [3.49] the wave forms [3.40], [3.41] and [3.42] for the displacements and we calculate the integrals over the period $2\pi/\tilde{\omega}$ of the introduced energy fluxes. It can be shown that, due to conservation of energy, these integrals do not depend on the space variable $\tilde{x}_1$. Owing to this independence on the variable $\tilde{x}_1$, in order to simplify calculations, we can simply introduce these integrals as:

$$J_i^l = \int_0^T \tilde{H}_i^l(0,\tilde{t})d\tilde{t}, \quad J_r^l = \int_0^T \tilde{H}_i^l(-\infty,\tilde{t})d\tilde{t}, \quad J_t^l = \int_0^T \tilde{H}_i^l(+\infty,\tilde{t})d\tilde{t}.$$

We are then finally able to introduce the reflection and transmission coefficients for longitudinal waves as:

$$R_l := \frac{J_r^l}{J_i^l}, \qquad T_l := \frac{J_t^l}{J_i^l},$$

which are such that $R_l + T_l = 1$. The reflection and transmission coefficients R_t and T_t for transversal waves are formally analogous.

We show in the following figures the behavior of reflection and transmission coefficients in terms of both the second gradient parameter ϵ and the frequency ω. These figures refer to both the longitudinal and transversal case: we have just to interpret ϵ as ϵ_l and ϵ_t, respectively.

Referring to Figure 3.7, we can start noticing that, when the second gradient parameter tends to zero, the generalized internal hinge (continuity of displacement and arbitrary normal derivative of displacement at the considered surface of discontinuity) implies that all the energy is transmitted independently of the value of the frequency. If the second gradient parameter is not vanishing, then the value of the frequency starts to play a role on the amount of energy which is transmitted or reflected so accounting for the description of dispersion phenomena. In particular, for very low frequencies (approaching to zero), the energy of the traveling wave continues to be almost completely transmitted independently of the value of the second gradient parameter ϵ^2. When increasing the frequency, the fact that the normal derivative of displacement at the discontinuity can take arbitrary values on both sides starts to play a role in the sense that the amount of reflected energy starts to increase, while the amount of transmitted energy decreases. This behavior can be directly related with the fact that, as is well known (see [MIN 65, TOU 64, TOU 62]), second gradient theories take into account the existence of an underlying microstructure even if remaining in the framework of a macroscopic model. More particularly, the fact that, for a given value of the second gradient parameter, an increasing frequency implies an increased amount of reflected energy can be explained if we think that the wavelength of the traveling wave decreases and becomes comparable to the characteristic length of the heterogeneities that are present in the material at a microscopic level. In other words, for very large wavelengths (small frequencies), the medium can be reasonably considered homogeneous even very close to the interface of discontinuity. In this case, when the considered wave reaches the interface, the continuity of displacements guarantees that the wave can continue undisturbed its path across the discontinuity itself. However, for smaller wavelengths (higher frequencies), the medium can no longer be considered homogeneous since the wave starts interacting with the underlying microscopic heterogeneities. More precisely, high frequency waves activate long range interactions at the microscopic level which result in an increasing amount of energy transported in the bulk because of the time derivative of displacement gradient (see equation [3.49] and subsequent considerations or, more generally, equation [3.20]). However, normal strains (normal gradient of displacement) at the two sides of the discontinuity interface are uncoupled because of the chosen boundary condition (generalized internal hinge) so that this amount of energy propagating because of microscopic interactions is actually reflected at the interface. The two just considered circumstances can be physically interpreted as follows: high frequency waves propagating in the bulk activate long-range interactions at a microscopic level and induce coupled deformations of close microscopic deformable structures, with a non-negligible amount of energy traveling because of this coupling. However, since at the considered interface, we assume that close microscopic structures situated at different sides are uncoupled, the amount of energy transported by these microscopically-interacting structures can only be reflected. Referring to Figure 3.7, we can finally remark that, for a fixed value of the frequency (and hence of the wavelength $1/k_l$), the amount of reflected energy increases for increasing values of the second gradient parameter up to the value $\epsilon^2 \approx 4$ and then it takes an almost constant value for greater values of ϵ^2. This is completely sensible and it means that, for a fixed value of the wavelength $(1/k_l)$, the amount of reflected energy increases with the value of the second gradient parameter ($\epsilon = L_l k_l$) in the range of values in which ϵ is sufficiently close to one (or equivalently in the range in which the considered wavelength is comparable to the second gradient characteristic length L_l). In this range, the amount of reflected energy becomes increasingly important when ϵ increases since the considered wavelength becomes comparable to the value of the second gradient characteristic length. For values of the second gradient parameter larger than four (or equivalently $L_l > 4/k_l$), the amount of reflected (transmitted) energy has a less relevant variation.

Considerations for the generalized internal roller are completely reversed with respect to those just made for the generalized internal hinge and can be deduced from Figure 3.7 reversing the curves of reflection and transmission coefficients. We start by noticing that when the second gradient parameter tends to zero, all the energy is reflected at the considered interface. Moreover, we notice that for very small values of the frequency (approaching to zero), approximately the total amount of the energy is still reflected independently on the value of the second gradient parameter. As before, this can be explained with the fact that, as the wavelength is sufficiently large then the medium can be considered homogeneous, and that the considered constraint allows arbitrary displacements on both sides of the considered interface. In some sense, it is like the interface was a free interface (no other medium on the other side) so that the incident wave can be reflected only. However, for non-vanishing values of the second gradient parameter and for increasing values of the frequency, some of the incident energy starts to be transmitted. As before, this can be explained with the fact that, for decreasing wavelengths, the wave starts interacting with the microstructure on the other side of the discontinuity surface. This means that the heterogeneities start moving at the microscopic level and create some mechanical interactions between the two sides of the considered interface, so allowing some energy transmission. The fact that for a fixed frequency, the generalized roller allows for an increasing transmitted energy when increasing the value of the second gradient parameter up to the value $\epsilon^2 \approx 4$ can be as before explained by considering that the second gradient interactions become more important when the wavelength becomes comparable to the characteristic second gradient length.

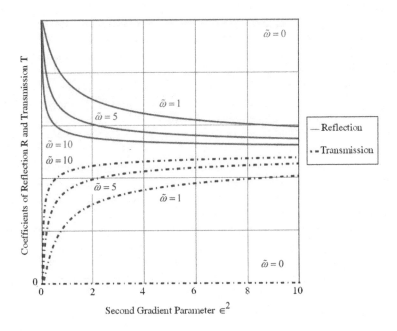

Figure 3.7. *Reflection and transmission coefficients for a generalized internal hinge $(k_S^n \to \infty)$ as a function of the second gradient parameter ϵ^2 and for different values of the non-dimensional frequency. The plots for the generalized internal roller are completely specular: the blue dashed lines must be referred to the reflection coefficient, while the red lines must be referred to the transmission coefficient*

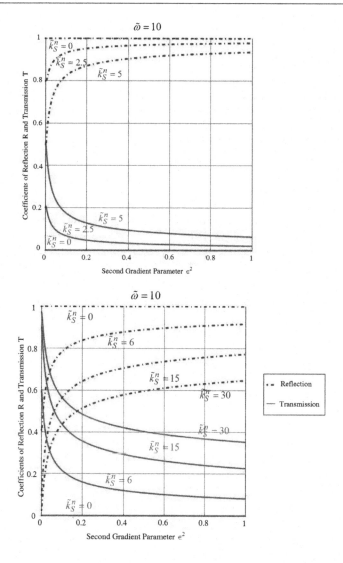

Figure 3.8. *Reflection and transmission coefficients for a generalized elastic internal hinge as a function of the second gradient parameter ϵ^2, for different values of the surface rigidity $k_S^n \in [0, 30]$ and for a fixed value of the non-dimensional frequency $\tilde{\omega} = 10$*

Figure 3.8 shows reflection and transmission coefficients for a generalized elastic internal hinge as a function of the second gradient parameter, for a fixed value of the frequency and for different values of the elastic rigidity at the interface. As we already noted, this constraint reduces to the generalized internal hinge when the rigidity $\tilde{k}_S^n$

tends to infinity. It can be remarked from this figure that when the elastic rigidity tends to zero, then this constraint reduces to the generalized internal roller. We can notice that a critical value of the elastic rigidity $\tilde{k}_S^n = 5$ exists corresponding to which a switch can be observed. More precisely, for values of the rigidity between 0 and 5, the larger amount of energy is reflected so that the considered constraint basically behave like a generalized internal roller. For values of the rigidity bigger than 5, a critical value of the second gradient parameter exists such that for ϵ^2 smaller than this critical value, the larger amount of energy is transmitted (the constraint behave as a generalized internal roller), while for ϵ^2 bigger than this critical value, the larger amount of energy is reflected (the constraint behave as a generalized internal hinge).

The presented results show that it is conceivable to go towards the development of suitable methods for getting estimates of some second gradient elastic moduli. Indeed, we propose to use a discontinuity surface between two macroscopically homogeneous second gradient solids to perform measurements of reflection and transmission coefficients at this interface: the presented results seem to supply a method for measuring the bulk properties of considered continua starting from these coefficients. The concepts presented here may be of interest as the boundary layer which may arise greatly influences wave transmission and reflection even in the considered simple cases. We assume that the discontinuity surface is the only possible localization of displacement (and normal gradient of displacement) eventual jumps. Moreover, we also consider interfaces on which surface elastic energy is concentrated. More complicated discontinuity surfaces may be conceived carrying mechanical properties which may directly influence transmission and reflection coefficients. Even if the investigation of the behavior of such more general discontinuity surfaces is of importance, the simplest discontinuity we considered here seems suitable for the conception of an effective experimentation. In a published paper (see [PLA 13]), we discussed how the fact that the interface possesses its own mass and elasticity influences the overall phenomenon of reflection and transmission. For the sake of conciseness, we do not report here the details concerning those results, but we limit ourselves to say that the aforementioned study allowed us to understand that (1) the presence of the interface has a considerable effect on wave reflection and transmission, above all for what concerns the presence of a distributed mass on the interface and (2) the effects related to the material interface on wave reflection and transmission can be dissociated from the effects due to the presence of microstructure (second gradient effects). We refer to the quoted paper above for additional details on these aspects.

3.3. Conclusions

We have shown in this chapter that generalized continuum models can be useful for the description of the dynamical behavior of microstructured materials. More particularly, we have shown in detail that:

– a relaxed micromorphic model is able to describe the onset of frequency band gaps, even when remaining in the simple linear-elastic, isotropic case. This means that such a generalized model is able to describe those situations in which macroscopic propagation of bulk waves is inhibited due to the presence of resonant microstructures or of microstructure-related diffusive phenomena;

– more generically, dispersive behaviors can be described by means of second gradient theories, even if band gaps cannot be accounted for with such models. We present in the second part of this chapter a systematic description of plane wave reflection and transmission at surfaces of discontinuity in second gradient materials. Such modeling of reflection and transmission phenomena might be of interest for the conception wave screens and wave absorbers.

Future studies should account for the possibility of including the presence of discontinuity surfaces in relaxed micromorphic media and of studying reflection and transmission at such surfaces. This would allow the study of complex systems, such as, for example, layered composites in which propagation is inhibited or permitted depending on the material properties and microstructures of the considered layer. Such layered materials would be useful, for example, for the conception and optimization of waveguides or of complex structures which are able to resist to seismic waves, or even to simple elastic vibrations.

4

Remodeling of Bone Reconstructed with Bio-resorbable Materials

Bone adaptation and remodeling is an amazing subject which has stimulated the minds of biologists and mechanicians ever since 1892 when Wolff (see [WOL 86]) observed that "internal architecture and external conformation of bones change in accordance with mathematical laws". Indeed, it is well established in scientific literature that interactions between mechanics and biology are crucial to correctly interpret and describe the behavior of growing tissues. This is strictly related to the fact that nature has developed "optimization methods" which, given the applied external loads, allow us to obtain a proper resistance against mechanical failure with a minimum use of material. Since the external applied loads unceasingly vary during life, living tissues must continuously be resorbed and synthesized in order to be able to resist the actual loads with the minimum possible quantity of matter. The functional adaptation of bone to mechanical usage implies the existence of a physiological control process. Essential components for the control process include sensors (osteocytes) for detecting mechanical usage and transducers to convert the usage measures into cellular responses (osteoblasts which synthesize new bone tissue and osteoclasts that resorb the old one). The cellular responses lead to gradual changes in bone shape and/or material properties and, once the structure has adapted sufficiently, the feedback signal is diminished and further changes to shape and properties are stopped. Although the presented description of biological phenomena occurring in growing tissues is certainly incomplete, it includes a number of processes which are definitely relevant when the observed mechanical adaptation takes place. A schematic of this feedback control process is shown in Figure 4.1 and, even if simplified, it captures the key features of the process at the macroscopic scale (millimeter or more).

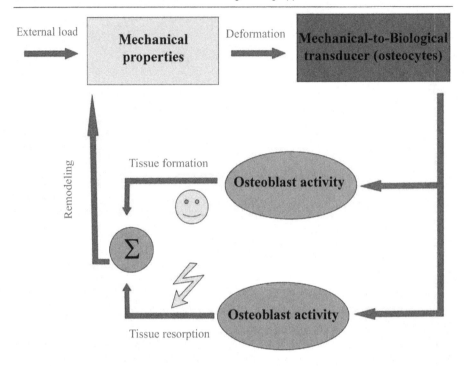

Figure 4.1. *Schematic representation of bone remodeling process*

Bone is a living organ that is constantly remodeled. It permanently replaces old worn tissue by newly formed bone, thereby avoiding impairment of the bone structure by wear and tear. It repairs damage and adjusts its structure to external stress. This directional self-renewal is one of the main advantages of the living bone tissue over any inanimate biomaterial, which is by definition subjected to deterioration. Thus, after an initial sustaining function immediately after surgery, an ideal bioactive material for bone substitution should be able to be completely resorbed and replaced by new bone. Bioresorbable bone fillers which are efficiently replaced by new bone tissue could provide a valid alternative to autografts, allografts and xenograft used in current bone grafting procedures to repair defects caused by surgery, tumors, trauma, implant revisions and infections and also for joint fusion. Surgeons currently use today scaffolds of synthetic or natural biomaterials that promote the migration, proliferation and differentiation of bone cells, but the success of these materials in repairing bone defects is, up to now, relatively limited. Indeed, bioresorbable grafts which are completely resorbed in the long term and replaced by new-formed bone tissue are not yet available due to the difficulty of fully understanding the complex process of reconstructed bone remodeling. Still, today, it is not rare to deal with failed surgeries of augmentation and replacement of harmed bone with bioresorbable materials and this clearly results in a heavy economic impact on the public health system. It is then clear that the optimization of bioresorbable scaffolds is nowadays one of the biggest challenges for tissue engineering which deserves the appropriate engagement and collaboration of biologists and mechanicians.

For all these reasons, biologists and mechanicians have to try today to make a point on which is the best way of approaching the problem of the filling of big bone defects by means of performant biomaterials which can be resorbed in the long run and replaced by newly formed bone tissue. As a matter of fact, a wealthy of conceptual difficulties arise when trying to make a bridge between the biological processes at the cellular scale and the need of designing performant bioresorbable grafts which have macroscopic extensions. We try in the following (non-exhaustive) list to make a point about the main difficulties which may arise when trying to build a theoretical bridge between the biomechanics of the macroscopic graft and the phenomena occurring at the cellular level:

– if we want to be able to predict the evolution in time of the whole implanted graft, then the most suitable modeling tools are continuum theories with the parallel use of finite element numerical methods. Such models are intrinsically macroscopic, in the sense that the representative elementary volume (REV) has a characteristic size of almost half a millimeter. This is equivalent to saying that, in a continuum theory, a small cube whose side is half a millimeter actually represents a material point which is the smallest entity which is seen by the model itself and of which we can follow the evolution. In other words, in a continuum model, we can think to follow the evolution in time of both bone and biomaterial, but it is not possible to directly follow informations concerning what happens in detail at the cellular scale;

– connected to the previous point is the challenge of a positive interaction between biologists and mechanicians in the view of the conception and realization of intelligent macroscopic structures which are able to be resorbed and replaced by new-formed bone. In order to successfully confront this challenge, the main problem to be solved is that of creating big pieces of artificial resorbable tissues in which cells can be implanted and which have to be able to survive in order to guarantee their activation which would finally lead to the resorption of the biomaterial and the simultaneous synthesis of new bone. The principal difficulty here is given by the extended size of the graft and the contemporary need of keeping cells alive inside the graft itself in order to gradually replace it by newly synthesized bone. It is for this reason that smart biomaterials have to be conceived which integrate the ability of furnishing the needed nutriment to the cells.

It is then clear that the problem of being able to produce extended "intelligent grafts" has to be mandatorily approached in the next years in order to go toward the possibility of setting up suitable finite element codes for the design of bioresorbable grafts for the replacement of big bone defects.

4.1. Generalized continuum theories for bone remodeling in the presence or absence of biomaterial

As has been noted at the beginning of this chapter, the process of bone remodeling as observed at the macroscopic scale consists of a strategic rearrangement of matter in such a way that the bone itself is able to resist to the externally applied loads in an optimized way. Such redistribution of matter is operated by specific cells at the lower scales whose joint activity allows us to appreciate such macroscopic remodeling as driven by the externally applied loads. It is then clear that the evolving microstructure of bone is a key factor to be accounted for in order to set up a macroscopic model which is able to describe remodeling phenomena in natural bone and, as it will be better pointed out in the following, also in artificially grafted tissues. One way to account for the effect of such evolving microstructures on the overall behavior of the considered living materials is that of considering one of the generalized continuum models introduced in the first chapter of this book. In particular, we will show in the remainder of this chapter how an *internal variable model* (see, for example, Figure 1.2) is able to account for the remodeling of both bone and biomaterial at the macroscopic scale and in a rather simplified way. We start by recalling that an internal variable model is one of the many possible generalized continuum models whose main characteristics are that of:

– having an extended kinematics with additional degrees of freedom (macroscopic displacement + extra degrees of freedom accounting for the evolution of the microstructure);

– having a strain energy density (SED) whose dependence on the introduced additional degrees of freedom does not involve space derivatives of such extra degrees of freedom (DOF). In other words, if we denote by u the macroscopic displacement field of the considered continuum and by P the set of its additional degrees of freedom, the SED for an internal variable model is of the type $W = W(\nabla u, P)$.

We will show in detail in the remainder of this chapter that if we limit ourselves to the introduction of two extra scalar degrees of freedom ρ_b^* and ρ_m^* representing the bone and biomaterial mass density, respectively, then the associated internal variable model is a reasonably simplified model which allows at the same time to:

– account for the macroscopic deformation of the bone-biomaterial system;

– account for the time evolution of bone and biomaterial mass densities at the macroscopic scale.

It is evident that a model of this type is intrinsically macroscopic since the informations that it is able to provide concern the deformation of relatively extended bone-biomaterial specimens together with the evolution in time of the bone-biomaterial densities through the specimen itself. Nevertheless, fundamental informations concerning the microstructure of the considered system are implicitly embedded in the presented internal variable model. Indeed, the time evolution of bone biomaterial densities as driven by externally applied loads is able to account for the presence of underlying cell activities taking place at the lower scales but which have evident macroscopic effects.

A continuum model of this type is a mandatory step if we want to reach the final goal of having performant computer design tools for the conception of extended bioresorbable grafts. In this optic, the interaction between the mechanicians and the biologists needs to be coordinated in order to go toward the conception of such intelligent macroscopic grafts which have the two main characteristics of:

– being extended enough to provide suitable answers to the problem of treating big bone defects;

– being suitably optimized (architecture of the microstructure, porosity, presence of nutriment, etc.) to guarantee the diffusion and survival of the needed cells inside the graft, even when natural bone and blood vessels are not yet present.

4.1.1. *Microstructure-induced deformation patterns in bone*

In this section, we make a brief review about the fact that bone is a hierarchically heterogeneous material and that the heterogeneity of its microstructures can sometimes lead to macroscopic deformation patterns which are not observable in more homogeneous materials. The question which we want to address here is not directly connected to the process of reconstructed bone remodeling that we want to analyze, but is related to a reflection concerning the influence of bone microstructure on its macroscopic deformation patterns. The readers who are specifically interested in the description of remodeling processes can hence skip this section, being it devoted to some general considerations about the effect of microstructure on the macroscopic mechanical response of bone.

As a matter of fact, when considering living microstructured materials such as natural bone tissue and its interactions with artificial bioresorbable materials, three fundamental questions naturally arise:

– Do natural living metamaterials (like bone) show microstructure-related macroscopic deformation patterns?

– If so, do these mechanical macroscopic manifestations of the microstructure affect certain biological phenomena which allow us to remodel (regenerate or resorb) such living tissues?

– How does the presence of an underlying microstructure affect the interactions between living bone tissues and bioresorbable grafts?

In the present stage of knowledge, it is actually possible to give a complete answer to the first question, but it is rather difficult to do the same for the remaining ones. In fact, if certainly the presence of high stress and strain concentrations must somehow affect the remodeling process, this is not the first problem to be treated at the present date. Indeed, it is much more urgent to treat in an efficient way the problem of the description of remodeling for simple situations not accounting for sharp strain variations which is what we do in the next sections. Nevertheless, we want to point out here that such particular microstructure-related macroscopic deformation patterns may be likely encountered in bone, even if the problem of precisely analyzing their effects on the process of remodeling must be postponed. We try in what follows to deal with the first question by presenting convincing arguments. We will then pass to the other questions and will try to understand why it is so difficult to construct a completely satisfying answer for them.

As the readers may have guessed, the answer to the first question is: yes, natural bone tissue evidently shows characteristic macroscopic deformation patterns which are due to the presence of the underlying microstructure. Such evidence has already been validated on experimental grounds. Indeed, several authors (see, for example, [BUE 03, LAK 82, YAN 82, YAN 81]) have found that the first level of microstructure, that of osteons, has some macroscopic manifestations when considering sufficiently small samples of bones. More particularly, it must be considered that bone is a hierarchically heterogeneous material, i.e. it can be considered as homogeneous at the scale of the millimeter (or more), but it starts presenting heterogeneities at the scale of the micron. At this scale, quasi-periodic circular structures (osteons) can be detected which confer highly heterogeneous properties to the material itself.

The mechanical effect of the presence of this first level of microstructure becomes increasingly evident on mechanical testing when considering specimens which are of the order of some millimeters. One possible way of dealing with such macroscopic manifestations of the microstructure of bone is to use particular generalized continuum theories (second gradient or micromorphic). Figure 4.2 shows the experimental points corresponding to torsion tests on specimens of bone with different characteristic size d. It can be noted, that in such case, Cauchy continuum theory (dashed line) does not allow us to describe the correct behavior of bone at small scales when considering sufficiently small specimens (d of the order of $10\ mm$ or less). However, when an enhanced Cosserat-type continuum theory is used (continuous line), the fitting with the experimental points becomes much more precise. The experimental test presented in Figure 4.2 can be interpreted by thinking that when a small specimen of bone is subjected to torsion, then the osteons start to rotate and deform of a different amount with respect to each other. Such local differential deformation can be explained by thinking that the local mechanical properties of single osteons are very heterogeneous, i.e. a given osteon can have very different stiffness with respect to the surrounding ones (see [HOC 06]). Hence, we can conclude that when considering a sample of bone of 10 mm or less (which is equivalent to consider sets of almost 200 osteons or less), then the deformation of the overall sample is affected by the differential deformation of the single osteons. The effect of microstructure on the global deformation of the sample is much more evident for specimens which contain from 6 to 60 osteons (i.e. specimens with $d \in [0.3, 3]\, mm$). When considering smaller specimens, clearly the mechanical properties of the sample tend to those of a single osteon, while when considering bigger specimens the effect of the heterogeneity of microstructure on the global mechanical properties of bone becomes negligible. Similar results have also been obtained by considering other loading conditions such as bending (see, for example, [YAN 82]).

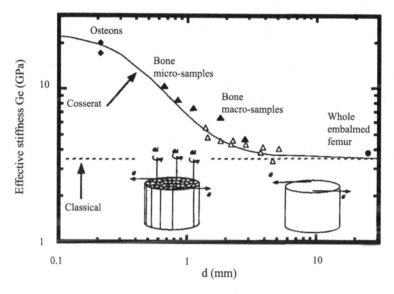

Figure 4.2. *Shear modulus versus size of the bone specimen (from [BUE 03])*

In the light of the aforementioned comments, it can be accepted that bone is a microstructured material which can be modeled by means of generalized continuum theories to correctly describe its mechanical behavior at the scale of some millimeters. This implies that a generalized continuum modeling can be useful to model such a living tissue when also considering bigger specimens (of the size of entire real bones), since the combination of particular loading and boundary conditions can activate the local deformation mechanisms defined above thus leading to stress and strain localization phenomena which would not be otherwise described by a classical Cauchy theory.

Once it has been accepted that the effect of microstructures on the global mechanical behavior of bone cannot always be neglected, the question which naturally arises is whether or not the observed microstructure-related deformation patterns affect the biological phenomena of bone remodeling. It is difficult to deal with this question for different reasons, some of which we try to summarize as follows:

– The complex biological phenomena which intervene in the process of bone remodeling are surely strongly connected to the mechanical state of bone (see, for example, [WOL 86]), but it is not completely clear which are the modalities according to which the biomechanical coupling takes place. More particularly, it is experimentally evident that new bone tissue is created by adapted cells (osteoblasts) when a certain bone region is highly deformed as the result of an external loading, but there is not an agreement on which is the trigger of such remodeling process. Some authors (see [HUI 00, MAD 11, MAD 12]) propose to use SED as such trigger: when the value of the SED exceeds a given threshold, then bone is too deformed and new bone synthesis takes place. The opposite mechanism is also currently observed, i.e. if bone is too compact it can be resorbed by other cells (osteoclasts) in order to be used in other places where it is really needed. Such a resorption phenomenon has been associated with the same authors to a value of the SED which is lower than the aforementioned threshold value. Nevertheless, in the literature, many other triggers have been suggested for the synthesis/resorption phenomena, including strain, (see [COW 76]) tissue damage (see [PRE 94]), daily stress stimulus (see [CAR 96]) and different forms of effective stress (see [DOB 02]). A deeper insight into the mechanically activated biology which is behind the remodeling process is needed at this stage in order to give a stable ground to mechanicians who want to provide suitable models for such phenomena. More particularly, this insight should be given on the macroscopic manifestation of the behavior of actor cells (osteoblasts and osteoclasts) and not on the behavior of single cells. What is more interesting to the eyes of a mechanician who wants to describe the effect of a macroscopic loading on a real bone is to understand how a cluster of cells reacts to a given mechanical environment: at which load level osteoblasts and osteoclasts are activated? Which is the (macroscopic) rate at which a suitable number of osteoblasts build a significative macroscopic volume, let us say $1\,mm^3$, of bone? Which is the (macroscopic) rate at which osteoblasts resorb the same volume of bone? Do the (macroscopic) rates of resorption and synthesis depend on the level of the applied load or are they constants? All such questions presently remain unclear and a huge, coordinated, effort from both mechanicians and biologists is needed in order to progress toward models which are increasingly realistic and predictive. The difficulty of answering to all the quoted questions is clearly related, among others, to the fact that it is very complicated to perform reliable and precise measurements on living organisms.

– Since the basic biomechanical mechanisms related to bone remodeling are not completely clarified at the present stage of knowledge, it is very complicated to say if, and to which extent, the fact that the material bone actually possesses a microstructure actually influences the process of bone remodeling. We believe that the presence of microstructure must necessarily influence such remodeling process since it is not uncommon to observe strain localizations (due, for example, to microscopic cracking) as a response to particular loading conditions. Such regions in which strain is highly localized are sensible to be regenerated even if the surrounding bone is much less deformed. A macroscopic model which wants to account for such effects of microstructure on remodeling of bones can be based on second gradient or micromorphic theories. Such models could be of use when introducing discontinuities in the natural bone tissue (for example, for a bone-graft implantation), when concentrations of stress and strain may be localized correspondingly to the interface separating the natural and the artificial tissue.

In conclusion, even if there is evidence that the remodeling phenomena in bone are activated, to a great extent, by mechanical excitation and that the presence of microstructure may somehow influence such remodeling processes, a considerable work still needs to be done in order to conceive second gradient or micromorphic models which may be of use in cases of real interest. A deeper level of scientific investigation is needed to relate more closely to the macroscopic effects that osteoblasts and osteoclasts have on sufficiently large samples of bone with the mechanical state of the samples themselves.

In virtue of the previous discussions, we are aware of the uncertainties which persist about the effect of concentration of stress and strains on the process of bone remodeling. It is then easy to understand that such uncertainties will be equally present when we want to deal with the macroscopic investigation of remodeling of bone in the presence of so-called bioresorbable materials when subjected to such heterogeneous deformation patterns. Over the last few decades, we have assisted in the broadening of the production and current use in medical practice of artificial bioresorbable materials which initially have the function of sustaining external mechanical loads, but that are progressively resorbed by osteoclasts and partially or completely replaced by natural bone tissue. These materials are conceived in order to have mechanical properties and microstructures that are the closest possible to those of natural bone. All the remarks which have been made in the case of natural bone tissue still remain valid when considering the coexistence of a natural and an artificial microstructured, resorbable material. It is not worth mentioning that other additional modeling difficulties arise in this case concerning, for example the possible interactions between the two materials, the migration and distribution of cells inside the implanted artificial material and so on.

4.2. A continuum two-solid mixture model for reconstructed bone remodeling

It is nowadays well established in scientific literature that interactions between mechanics and biology are crucial to correctly interpret and describe the behavior of growing tissues (see, for example, [CAR 96, HUI 00, MAD 11, CAS 10]). This is strictly related to the fact that nature has developed optimization methods which, given the applied external loads, allow us to obtain a proper resistance against mechanical failure with a minimum use of material.

The amazing problem of bone adaptation and remodeling stimulated the minds of biologists and mechanicians ever since the last years of 1,800 when Wolff (see [WOL 86]) observed that "internal architecture and external conformation of bones changes in accordance with mathematical laws". The 20th Century has then seen an explosion of the research in the field and different mathematical models have been proposed to describe functional adaptation and bone remodeling (see, for example, [HUI 00, RUI 05, WEI 92]).

Today, we are aware of the fact that the cells which are involved in functional adaptation of bone are divided into two main big classes:

– the sensor cells, (or osteocytes) which are able to detect the external mechanical stimulus and transduce it in a suitable biological signal which can be decoded by other cells;

– the actor cells (osteoblasts and osteoclasts) which detect the signal emitted by the osteocytes and, respectively, synthesize and resorb bone tissue depending on the state of mechanical excitation.

As already mentioned, various mechanical stimuli have been proposed as triggers for bone adaptation, including strain, (see [COW 76]) SED (see, for example, [HUI 00]), tissue damage (see, for example, [PRE 94]), daily stress stimulus (see [CAR 96]) and different forms of effective stress (see [DOB 02]). In this chapter, we follow the idea of adopting the SED as the principal trigger of bone remodeling. More precisely, we assume that the osteocytes are well placed within bone porosity to function as "strain gauges", and to emit a signal (stimulus) the intensity of which is proportional to the measured strain energy and to bone apparent density. Moreover, we assume that a threshold value of the stimulus exists such that osteoblasts (bone synthesis) are activated if the value of stimulus is higher than this threshold, while osteoclasts (bone resorption) are activated if the value of stimulus is lower than the threshold itself. The hypothesis of existence of such a threshold value is based on the idea that a high value of the deformation energy is associated with a need of a more compact bone, while a low value of strain energy may be associated with a surplus of material at a given location which can then be resorbed and reused in other locations subjected to higher mechanical solicitations.

The understanding of the profound interactions between mechanics and biology in functional bone adaptation and remodeling naturally led to the use in medical practice of artificial bioresorbable scaffolds which initially have the function of sustaining external mechanical loads, but are progressively resorbed by osteoclasts and partially or completely replaced by natural bone tissue. While continuum models for the description of natural bone regeneration are widely spread in the scientific literature (see among many others [COW 76]), the conception of rigorous models allowing for the description of biomaterials resorption and of their gradual replacement by natural bone tissue is still an open challenge. We introduce a continuum mixture model which allows for describing phenomena of natural bone remodeling on one hand and, on the other hand, phenomena of resorption of bioresorbable artificial materials followed by a gradual substitution with natural bone tissue.

4.2.1. Equilibrium equations accounting for mass creation and dissolution as driven by biomechanical coupling

Following standard procedures of mixture theory, we describe the deformation of a two-solids continuum mixture by introducing a Lagrangian (or reference) configuration $B_L \subset \mathbb{R}^3$ and a suitably regular kinematical field $\chi(\mathbf{X}, t)$ which associates with any material point $\mathbf{X} \in B_L$ its current position $\mathbf{x}$ at time t. The kinematics of the system is completed by introducing two Lagrangian densities $\rho_b^*(\mathbf{X}, t)$ and $\rho_m^*(\mathbf{X}, t)$ which represent the density of the natural bone tissue and the artificial material, respectively. A generalized continuum model of this type has been introduced at the beginning of this chapter as "internal variable model". We explicitly note that, in our mixture model, we associate with each material particle $\mathbf{X}$ two different reference densities which can evolve in time. The image of the function χ gives at any instant t the current shape of the body $B_E(t)$: this time-varying domain is usually referred to as the Eulerian configuration of the mixture and it represents the system during its deformation. Since we will use it in the following, we also introduce the displacement field $\mathbf{u}(\mathbf{X}, t) := \chi(\mathbf{X}, t) - \mathbf{X}$, the tensor $\mathbf{F} := \nabla\chi$ and the Green–Lagrange deformation tensor $\mathbf{E} := (\mathbf{F}^T \cdot \mathbf{F} - \mathbf{I})/2$.

According to the fact that we are using an internal variable model, we introduce the SED $U^*(\mathbf{E}, \rho_b^*, \rho_m^*)$ which is assumed to depend on the Green–Lagrange deformation tensor $\mathbf{E}$ and also on the Lagrangian apparent mass densities (internal variables) of both bone and biomaterial. It can be shown, for example, by means of a suitable variational principle (see, for example,. the methods presented in [SCI 08]) that, neglecting external body forces, the bulk equilibrium equation for such a system reads:

$$ div \left[\mathbf{F} \cdot \frac{\partial U^*}{\partial \mathbf{E}} \right] = 0 \qquad\qquad [4.1] $$

together with the following duality conditions valid on any discontinuity surface $\Sigma \subset B_L$ of such a continuum:

$$\left[\!\!\left[\left(\mathbf{F} \cdot \frac{\partial U^*}{\partial \mathbf{E}} \cdot \mathbf{n}\right) \cdot \delta\mathbf{u}\right]\!\!\right] = 0, \qquad\qquad [4.2]$$

where $\mathbf{n}$ is the unit normal vector to the surface Σ and $\delta\mathbf{u}$ is the variation of the displacement field. Finally, given a quantity a defined everywhere and having continuous traces a^+ and a^- on the two sides of Σ, respectively, we have set $[\![a]\!] := a^+ - a^-$.

4.2.2. *Bone remodeling equations*

We now want to formulate a suitable evolutionary problem for the introduced kinematical fields which is able to catch the most important features of the remodeling processes occurring in bone tissue after initial healing and in the presence of bioresorbable grafts initially functioning as a bone tissue filler and support. Before introducing the differential equations which we believe to be suitable to accomplish this task, we recall here some basic biological facts which led us to use them in our modeling. We can start distinguishing two types of cells (which belong to the common class of so-called actor cells), namely the osteoblasts (specialized in new bone tissue formation) and the osteoclasts (which are able to resorb both natural bone and bioresorbable material). We assume that these two kinds of cells are present everywhere in both the living bone tissue and the artificial material under the unique condition that a suitable porosity is present. In other words, we are setting up a model which can be used to describe those situations in which:

– a suitable period of time has passed from the implantation of the graft and cells have migrated everywhere in the scaffold (this is a reasonable situation which is rapidly attainable for small grafts);

– the graft has been engineered in such a way that cells have been perfused inside the graft before its implantation and the needed nutriment is also provided for their survival (such intelligent grafts do not exist today yet, but an effort arising from the collaboration of biologists and mechanicians could bring to their conception in the next years).

The activity of osteoblasts and/or osteoclasts is regulated by the instructions of a signal generated by a third kind of cell called osteocytes: this signal is proportional to the deformation which the osteocytes can measure at a given point (for this reason, they are also called sensor cells). In order to measure deformation energy at a given point, the sensor cells do not move and spend all their life in that point. Sensor cells originate from osteoblasts when these latter have accomplished their task of synthesizing new bone around them: when an osteoblast is completely surrounded by new natural bone tissue, it changes its nature and becomes an osteocyte, i.e. a sensor cell which starts to measure deformation and emit a signal proportional to its measured value. This brief and simplified description of the biology which is known to take place behind the process of reconstructed bone remodeling is sufficient to justify the remodeling equations which we choose to use in this chapter. For a more detailed description of the biological phenomena occurring in natural bone and artificial bioresorbable material remodeling, we refer to [MAD 11, MAD 12]. Given the coupled biological and mechanical phenomena described avove, we are now able to postulate a proper set of differential equations which are able to catch the macroscopic effect of the activity of osteoblasts and osteoclasts on the process of remodeling. We choose the evolutionary equations for apparent densities to be simply first-order ordinary differential equations with respect to time. In formulas, we assume that:

$$\frac{\partial \rho_b^*}{\partial t} = \mathcal{A}_b(\varphi^*, S^*), \qquad \frac{\partial \rho_m^*}{\partial t} = \mathcal{A}_m(\varphi^*, S^*), \qquad [4.3]$$

where φ^* is the porosity of the considered continuum mixture and S^* is the biological stimulus which accounts for the activity of osteocytes. We will duly explain in the following how the porosity φ^* and the stimulus S^* are assumed to constitutively depend on the introduced basic kinematical fields.

The constitutive equations for $\mathcal{A}_b$ and $\mathcal{A}_m$ must account for different phenomena of biological and mechanical nature and for some geometrical circumstances, namely:

1) the surface available for resorption or synthesis inside the considered macroscopic elementary volume depends on the effective porosity field;

2) the different properties of bone tissue and biomaterial determine different resorption rates, due to the different effect of actor cells on these different materials;

3) a positive stimulus triggers synthesis of natural bone tissue, while a negative stimulus gives rise to resorption of both natural bone and biomaterial.

The previous considerations are consequence of the biological nature of the process of synthesis and resorption: in particular, osteoclasts adsorb at the same time (and presumably they are not able to be completely selective) both the bone tissue and the resorbable material, while osteoblasts produce only bone tissue. In this chapter, following what is done in [MAD 11, MAD 12], we will assume that:

$$\mathcal{A}_b(\varphi^*, S^*) = A_b(S^*)H(\varphi^*), \qquad \mathcal{A}_m(\varphi^*, S^*) = A_m(S^*)H(\varphi^*), \qquad [4.4]$$

where the functions A_b and A_m are piece-wise linear functions with different slopes for negative and positive values of stimulus (remark that, according to assumption (3), A_m must vanish when $S^* > 0$, as there cannot be synthesis of biomaterial) and the function H is designed in order to account for the influence of effective porosity on the biological activity of actor cells: when effective porosity is too large, there is not enough material on which actor cells may deposit, when it is too small there is not enough free space in the pores to allow their mobility and deposit. We choose the shape of H in such a way that $H = 0$ for $\varphi = 0$ or $\varphi = 1$. In particular, the following forms for the functions $H(\varphi)$, A_b and A_m have been chosen to perform numerical simulations:

$$H = k\,\varphi^*(1 - \varphi^*), \qquad A_b(S^*) = \begin{cases} s_b\,S^* & \text{for} \quad S^* > 0 \\ r_b\,S^* & \text{for} \quad S^* < 0 \end{cases},$$

$$A_m(S^*) = \begin{cases} 0 & \text{for} \quad S^* > 0 \\ r_m\,S^* & \text{for} \quad S^* < 0 \end{cases}, \qquad [4.5]$$

where k is a suitable constant parameter, s_b and r_b will be called synthesis rate and resorption rate for bone tissue, respectively, and r_m will be called resorption rate for biomaterial.

In order to close the problem, we need to explicitly give the constitutive relationships for the porosity and the stimulus as functions of the introduced kinematical fields. In what follows, we will assume that:

$$\varphi^* = 1 - \theta\,\frac{(\rho_b^* + \rho_m^*)}{\rho_{max}}, \qquad [4.6]$$

$$S^*(\mathbf{X}, t) = \left(\int_{B_L} U^*(\mathbf{X}_0, t)d^*(\mathbf{X}_0, t)\exp(-D\,\|\chi(\mathbf{X}) - \chi(\mathbf{X}_0)\|)d\mathbf{X}_0 \right) - S_0^*(\mathbf{X}, t),$$

$$d^* = \eta\frac{\rho_b^*}{\rho_{max}}, \qquad [4.7]$$

where $\theta \in [0, 1]$, ρ_{max} is the density of compact bone (corresponding to minimum porosity), d^* is the fraction of osteocytes (it is assumed to be proportional to ρ_b^* since osteocytes take birth from osteoblasts when they are completely surrounded by bone tissue, hence osteocytes can be present only simultaneously with natural bone), $1/D$ is a length accounting for the range of action of the signal sent by osteocytes and $\eta \in [0, 1]$. Moreover, S_0^* represents a threshold value for the stimulus which discriminates between resorption and synthesis: if the integral on the right-hand side of [4.7]$_1$ is smaller than S_0^*, then the stimulus is negative and resorption takes place, while if it is greater than S_0^* the overall stimulus is positive and synthesis of natural tissue occurs. The evolution equations [4.3], because of constitutive equations [4.4], are nonlinear and depend on the value of strain energy of the whole body by means of the integral operator [4.7] which is nonlinear as well. The stimulus introduced here describes only some precise characteristics of the network of sensor and actuator cells in living bone tissues since it has a rather complex (and not completely known) behavior. The mechanisms of signal transmission occurring in this network would deserve careful further studies.

4.3. A simple one-dimensional, linearized, isotropic problem

We consider here a simple one-dimensional (1D) linearized case which we can easily solve and which allows us to understand the potentialities of the proposed model in a simplified way. To do so, let us consider a Lagrangian reference frame (X, Y, Z). We assume that the displacement vector has only one non-vanishing components along the X axis, i.e. $\mathbf{u}(X, t) = (u(X, t), 0, 0)$. With this assumption and assuming the hypothesis of small displacements (linear case), it is easy to show that:

$$\mathbf{F} = \begin{pmatrix} (u'+1) & 0 & 0 \\ 0 & 1 & 0 \\ 0 & 0 & 1 \end{pmatrix}, \qquad \mathbf{E} = \frac{1}{2}\begin{pmatrix} u'(u'+2) & 0 & 0 \\ 0 & 0 & 0 \\ 0 & 0 & 0 \end{pmatrix} \Rightarrow \varepsilon = \begin{pmatrix} u' & 0 & 0 \\ 0 & 0 & 0 \\ 0 & 0 & 0 \end{pmatrix}. \quad [4.8]$$

where we denote by an apex the partial differentiation with respect to the space variable X and where ε is the linearized Green–Lagrange deformation tensor.

Let us also consider the following particular quadratic form for the SED valid for the isotropic case:

$$U^*(\varepsilon, \nabla\varepsilon, \rho_b^*, \rho_m^*) = \mu(\rho_b^*, \rho_m^*) \langle \varepsilon, \varepsilon \rangle + \frac{\lambda(\rho_b^*, \rho_m^*)}{2} (tr(\varepsilon))^2. \qquad [4.9]$$

Indeed, due to its hierarchical microstructure, bone often behaves as an anisotropic material, so that the considered expression for the deformation energy should be generalized to describe such more complicated cases. The parameters λ and μ are the classical Lamé parameters of the mixture which can also be given in terms of the Young's modulus E and the Poisson's coefficient ν.

In the considered 1D linearized case, the proposed expression of the strain energy simplifies into:

$$U^*(u', \rho_b^*, \rho_m^*) = \frac{1}{2} \frac{E(\rho_b^*, \rho_m^*)}{N} (u')^2, \qquad [4.10]$$

where E is the Young's modulus of the mixture, ν is its Poisson's coefficient and where we set , $N := (1+\nu)(1-2\nu)/(1-\nu)$.

In the considered example, the elasticity parameters are not constant as in classical elasticity, but vary (in space and time) as functions of the apparent mass densities

$\rho_b^*(\mathbf{X}, t)$ and $\rho_m^*(\mathbf{X}, t)$. In particular, we will assume in what follows that the Poisson's coefficient remains constant in space and time, while we consider the following forms for the Young's modulus as a function of the bone and biomaterial densities:

$$E(\rho_b^*, \rho_m^*) = \left(E_b \left(\frac{\rho_b^*}{\rho_{max}} \right)^\beta + E_m \left(\frac{\rho_m^*}{\rho_{max}} \right)^\beta \right), \qquad [4.11]$$

where all the quantities which have not been introduced before are constants. In particular, ρ_{max} is the maximum possible density attainable by natural bone and suitable values can be found in the literature for this parameter, while E_b and E_m can be thought as the constant Young's modulus of compact bone and biomaterial, respectively, (also such values can be easily fixed).

4.3.1. Mechanical equilibrium equation and naturally associated boundary conditions

In order to show the potentialities of the introduced model, we start considering the deformation of a 1D continuum of length L clamped at $X = 0$ and subjected to a simple tension $\mathbf{f}^{ext} = (f^{ext}, 0, 0)$ at $X = L$. At $X = L/2$ (in the middle of the specimen), we consider a surface of discontinuity of the material properties (interface between natural bone and artificial material at the initial time) such that the two considered materials are clamped together.

With the assumptions made for the considered 1D, isotropic case, and considering a constant Poisson coefficient (which is assumed to stay constant also during the remodeling) the bulk equilibrium equation [4.1] in the two considered subregions takes the form:

$$\left(E(\rho_b^*, \rho_m^*)\, u' \right)' = 0. \qquad [4.12]$$

An equation of this type governs the motion of the 1D continuum mixture both for $X \in [0, L/2)$ and $X \in [L/2, L]$. We remark that the 1D application presented here does not account for the presence of transversal sections in the considered specimen and only considers a simple tensile external load.

As far as jump conditions are concerned, they must be chosen on the basis of the physics of the problem to be treated. For the considered problem, we can say that the clamp at $X = 0$ results in the kinematical condition $\mathbf{u} = 0$; the internal clamp gives rise to the jump conditions $[|\mathbf{u}|] = 0$, $[|\mathbf{F} \cdot \partial U^* / \partial \mathbf{E} \cdot \mathbf{n}|] = 0$, in $X = L/2$; finally, at the free end $X = L$ the following condition holds $\mathbf{F} \cdot \partial U^* / \partial \mathbf{E} \cdot \mathbf{n} = f^{ext}$. It can be checked, recalling the considered isotropic expression for the deformation energy, that in the treated 1D, linearized case these boundary conditions simplify into:

$$u = 0, \qquad\qquad\qquad\qquad \text{in } X = 0,$$

$$[|u|] = 0, \qquad \left[\left|\frac{E(\rho_b^*, \rho_m^*)}{N} u'\right|\right] = 0, \qquad\qquad \text{in } X = L/2 \qquad\qquad [4.13]$$

$$E(\rho_b^*, \rho_m^*) \, u' = F^{ext}, \qquad\qquad\qquad \text{in } X = L,$$

where we set $F^{ext} = N f^{ext}$. We remark that for any given value of the densities ρ_b^* and ρ_m^*, equation [4.12] can be easily integrated once in both the regions $X > L/2$ and $X < L/2$ and that a solution for u' can be easily found by using only those boundary conditions [4.13] which involve the first derivatives of u.

We also finally note that in the considered 1D case, the expression [4.7] for the stimulus simplifies into:

$$S^*(X,t) = \left(\int_0^L U^*(X_0, t) \, d^*(X_0, t) \exp(-D \, \|\chi(X) - \chi(X_0)\|) dX_0\right) - S_0^*(X,t),$$

$$[4.14]$$

where we recall that the strain energy U^* and the osteocytes fraction d^* are constitutively given in equations [4.10] and [4.7]$_2$, respectively.

4.3.2. Non-dimensional form of mechanical and biological equations

4.3.2.1. Dimensionless mechanical equations and solution for the 1D strain

We now want to write all the equations in non-dimensional form and find an explicit solution for the strain. To do so, we start from mechanical equations and we introduce the non-dimensional variables $\tilde{x} = X/L$, $\tilde{u} = u/u_0$ and $\tilde{E} = E/E_b$, where L is a characteristic macroscopic length which can be chosen to be, for example, the length of the considered specimen, u_0 is a characteristic displacement, E_b is a constant elastic modulus which we choose to be that of compact bone (we choose the same characteristic quantities L, u_0 and E_b for the two regions $X > L/2$ and $X < L/2$). Using these new dimensionless quantities, the bulk equations [4.12] can be rewritten in the two regions as:

$$\left(\tilde{E}(\rho_b^*, \rho_m^*) \, \tilde{u}'\right)' = 0, \qquad\qquad\qquad [4.15]$$

where from now on the apex indicates, with a slight abuse of notation, the derivative with respect to the dimensionless space variable $\tilde{x}$. Analogously, the boundary conditions [4.13] can be rewritten as:

$$\tilde{u} = 0, \qquad\qquad\qquad \text{in } \tilde{x} = 0,$$

$$[|\tilde{u}|] = 0, \qquad \left[\left|\frac{\tilde{E}(\rho_b^*, \rho_m^*)}{N}\, \tilde{u}'\right|\right] = 0, \qquad \text{in } \tilde{x} = 1/2 \qquad [4.16]$$

$$\tilde{E}(\rho_b^*, \rho_m^*)\, \tilde{u}' = \tilde{F}^{ext}, \qquad\qquad \text{in } \tilde{x} = 1,$$

where we set $\tilde{F}^{ext} = L/(E_b u_0)F^{ext}$. We now explicitly state that equation [4.15] can be integrated once and an explicit solution for $\tilde{u}'$ can be found using those boundary conditions in [4.16] which involve the first derivatives of $\tilde{u}$. We can limit ourselves to find a solution for $\tilde{u}'$ instead of looking for the displacement $\tilde{u}$, since the former is the variable appearing in the SED [4.22] which is needed to calculate the biological stimulus. It is easy to check that such a solution takes the form:

$$(\tilde{u}')^- = \frac{N^-}{\tilde{E}^- N^+}\tilde{F}^{ext}, \qquad (\tilde{u}')^+ = \frac{1}{\tilde{E}^+}\tilde{F}^{ext}, \qquad [4.17]$$

where from now on, when it is needed to be explicitly specified, we will label by a $-$ and a $+$ superscript the quantities defined on the regions $\tilde{x} < 1/2$ and $\tilde{x} > 1/2$, respectively.

4.3.2.2. *Dimensionless evolution equations for the densities*

We also want to find a dimensionless form of the biological evolution equations: dividing equations [4.3] by ρ_{max}, introducing the dimensionless time $\tilde{t} = t/t_0$ and considering the constitutive assumptions [4.5], their dimensionless form reads:

$$\frac{\partial}{\partial \tilde{t}}\left(\frac{\rho_b^*}{\rho_{max}}\right) = H(\varphi^*)\tilde{A}_b(\tilde{S}^*), \qquad \frac{\partial}{\partial \tilde{t}}\left(\frac{\rho_m^*}{\rho_{max}}\right) = H(\varphi^*)\tilde{A}_m(\tilde{S}^*). \quad [4.18]$$

In these equations, we set:

$$H = k\,\varphi^*(1-\varphi^*), \qquad \tilde{A}_b(\tilde{S}^*) = \begin{cases} \tilde{s}_b\,\tilde{S}^* & \text{for} \quad \tilde{S}^* > 0 \\ \tilde{r}_b\,\tilde{S}^* & \text{for} \quad \tilde{S}^* < 0 \end{cases},$$

$$\tilde{A}_m(S) = \begin{cases} 0 & \text{for} \quad \tilde{S}^* > 0 \\ \tilde{r}_m\,\tilde{S}^* & \text{for} \quad \tilde{S}^* < 0 \end{cases}, \qquad\qquad [4.19]$$

where φ^* is still given by equation [4.6] and moreover:

$$\tilde{s}_b = \frac{U_0 L t_0}{\rho_{max}} s_b, \qquad \tilde{r}_b = \frac{U_0 L t_0}{\rho_{max}} r_b, \qquad \tilde{r}_m = \frac{U_0 L t_0}{\rho_{max}} r_m$$

$$U_0 = \frac{E_b}{N} \left(\frac{u_0}{L} \right)^2 . \tag{4.20}$$

Finally, the dimensionless Lagrangian stimulus $\tilde{S}^*$ appearing in [4.19] is:

$$\tilde{S}^*(\tilde{x}, \tilde{t}) = \frac{S^*}{U_0 L} = \left(\int_0^1 \tilde{U}^*(\tilde{x}_0, \tilde{t}) \, d^*(\tilde{x}_0, t) \right.$$

$$\left. \exp \left(-\tilde{D} \left\| \tilde{x} - \tilde{x}_0 + \frac{u_0}{L} (\tilde{u}(L\tilde{x}) - \tilde{u}(L\tilde{x}_0)) \right\| \right) d\tilde{x}_0 \right) - \tilde{S}_0^*(\tilde{x}, \tilde{t}), \quad [4.21]$$

where:

$$\tilde{U}^* = \frac{U^*}{U_0} = \frac{1}{2} \tilde{E} \, (\tilde{u}')^2 , \tag{4.22}$$

and where we introduced the new dimensionless quantities $\tilde{D} = DL$, $\tilde{x}_0 = X_0/L$ and $\tilde{S}_0^* = S_0^*/(U_0 L)$. Note that, since in linear elasticity the quantity u_0/L is often very small, the third term inside the norm can be neglected in most of the practical applications.

We also remark for completeness that, in virtue of equation [4.11], the dimensionless evolving Young's modulus takes the form:

$$\tilde{E}(\rho_b^*, \rho_m^*) = \left(\left(\frac{\rho_b^*}{\rho_{max}} \right)^\beta + \frac{E_m}{E_b} \left(\frac{\rho_m^*}{\rho_{max}} \right)^\beta \right) . \tag{4.23}$$

4.3.2.3. Numerical resolution algorithm

Summarizing, we can say that the problem which will be treated in the following numerical simulations will consist of the following steps:

1) Fix the constitutive values of the parameters of the proposed model (see, for example, Tables 4.1 and 4.2).

2) Fix the initial values for the mass densities ρ_b^*/ρ_{max} and ρ_m^*/ρ_{max} (the initial value of $\tilde{E}$ is fixed as well in virtue of equation [4.23]).

3) Solve the differential equations [4.15] equipped with the boundary conditions [4.16], or equivalently use directly the solution [4.17] to determine $\tilde{u}'$.

4) Calculate the SED by means of equation [4.22] and the fraction of osteocytes by using equation [4.7]$_2$.

5) Calculate the biological stimulus by means of equation [4.21].

6) Calculate the mass densities ρ_b^*/ρ_{max} and ρ_m^*/ρ_{max} at the subsequent time step by means of equations [4.18].

7) Return to point (3) and repeat up to convergence.

We explicitly remark that the introduced continuum mixture model allows for the description of changes in the distribution of the two solid phases present in reconstructed bone grafts.

4.4. Numerical simulations

We start by choosing the values of the constitutive parameters as shown in Tables 4.1 and 4.2.

ρ_b/ρ_{max} at $t = 0$	ρ_m/ρ_{max} at $t = 0$	k	θ	η	$\tilde{s}_b$	$\tilde{r}_b$	$\tilde{r}_m$
0.5	0.5	4	1	1	10	10	15

Table 4.1. *Values of the parameters used in numerical simulations*

$\tilde{F}^{ext}$	$\tilde{E}_b$	$\tilde{E}_m$	N^-	N^+	β	$\tilde{D}$	$\tilde{S}_0^*$
2.5	1	1	1	1	1.9	10	0.1

Table 4.2. *Values of the parameters used in numerical simulations*

Always with reference to Table 4.1, we are considering constant initial densities for the bone and the biomaterial: the initial distribution of such dimensionless densities which are used in our simulations is shown in Figure 4.3.

We explicitly remark that, since we are using a two-solid mixture model, the initial values of both the natural bone and biomaterial mass densities must be simultaneously specified at all points, even if one of the two is vanishing.

For the chosen values of the parameters, we look for the equilibrium configuration of the specimen by following the steps (1)–(7) detailed in the previous section. Such equilibrium configuration is shown in Figure 4.4.

By comparing Figures 4.3 and 4.4, we can note that creation of new bone tissue occurs in the region originally occupied by natural bone tissue (left half of the specimen) as a result of the application of the external mechanical excitation: this fact results in an increase in the bone apparent mass density which attains its maximum possible value in this region. As far as the region originally occupied by artificial bioresorbable material (right half of the specimen) is concerned, we can see how the application of the external mechanical load triggers the resorption of the biomaterial which is subsequently replaced by natural bone tissue. We can see in Figure 4.4 that, at the end of the process of remodeling, an extended region of a natural bone/artificial material composite still exists even relatively far from the initial discontinuity surface. The fact that the artificial material is not completely resorbed can be explained with the following mechanisms taking place during the remodeling process:

– the signal sent by the osteocytes (which are initially located only in the left half of the specimen) reaches the actor cells which are located in the artificial material and which are closer to the discontinuity surface. This signal is intense enough that no resorption of biomaterial takes place and natural bone tissue starts being created;

– this process of generating new natural tissue close to the interface is so rapid that the available porosity is quickly filled and no more space is available for the actor cells to deposit and continue synthesizing more natural tissue;

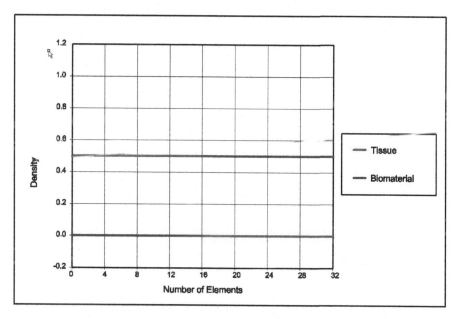

Figure 4.3. *Initial distribution of natural bone and artificial biomaterial mass densities in the considered specimen. For a color version of the figure see www.iste.co.uk/madeo/continuum.zip*

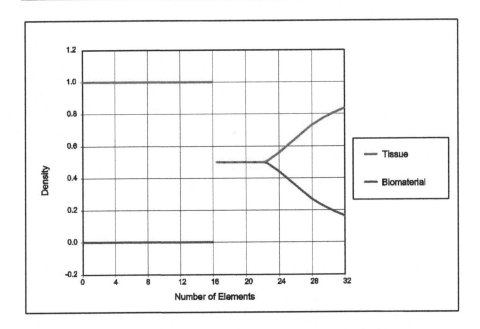

Figure 4.4. *Final distribution of natural bone and artificial biomaterial mass densities for the parameters shown in Tables 4.1 and 4.2. For a color version of the figure see www.iste.co.uk/madeo/continuum.zip*

– in regions of the artificial material which are far from the interface, the biological stimulus reaching the actor cells is lower than the chosen threshold value, so that resorption of the artificial biomaterial starts taking place. At the same time, since new natural tissue is forming close to the interface and starts propagating in the biomaterial region, an increasingly intense stimulus is perceived by actor cells which starts synthesizing natural bone tissue once the intensity of the stimulus overcomes the threshold value (i.e. S^* becomes positive);

– a more pronounced resorption occurs far from the interface since the osteoclasts have more time to resorb the artificial material before that the intensity of the stimulus overcomes the threshold value and that natural bone tissue synthesis occurs.

4.4.1. *Effect of applied external force and biomaterial Young's modulus on remodeling*

We start showing how the variation of the externally applied load influences the entity of the replacement of biomaterial with new-formed bone tissue.

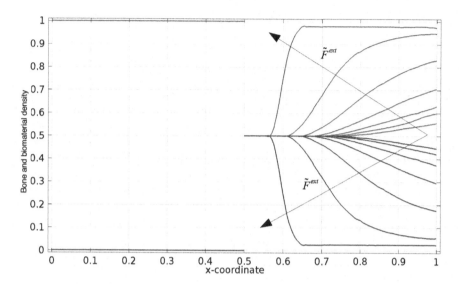

Figure 4.5. *Final distribution of natural bone and artificial biomaterial mass densities for different values of the dimensionless force $\tilde{F}^{ext} \in [1.6, 3.4]$. For a color version of the figure see www.iste.co.uk/madeo/continuum.zip*

Figure 4.5 shows that increasing the applied external force in a suitable range results in an increased percentage of replacement. This fact allows us to conclude that, once a suitable loading condition is found that allows for some replacement of the biomaterial with natural bone tissue, the intensity of such a load can be suitably tuned in order to produce an optimal replacement of the artificial graft.

We explicitly state here that in the proposed model we considered only the application of static external forces. We believe that such an approximation is rather reasonable since a mechanical dynamic load has characteristic times which are of the order of minutes, while the remodeling process takes place on characteristic times which are of the order of weeks. Nevertheless, we mention that some experimental studies suggest (see, for example, [GÓM 11, GON 10]) that the effect of frequency on the remodeling process is not negligible for reasons that still are to be investigated. In order to account for such phenomena, a more refined model with two intrinsic time-scales (one for the applied dynamical load and the other for the density variation) should be conceived in order to appreciate such differences. However, other studies on ultrasound treatment of bone defects indicate that the most important parameter influencing the bone regeneration is not the frequency but the intensity of the applied load (see, for example, [HAS 10, LAV 09]). This is in accordance with our model and supports on experimental grounds our simplifying assumption concerning the use of a unique weekly time-scale associated with the remodeling process.

We finally present the results of our model concerning the effect of the variation of the rigidity of the biomaterial constituting the graft. Indeed, such a parameter can be suitably tuned by choosing the elastic properties of the material in such a way to favor its resorption and replacement with new bone tissue. Figure 4.6 shows that the replacement of the graft is much higher when considering biomaterials that are 10 to 20% more rigid than the natural bone. Of course, such a result would be totally indicative only if rigid grafts optimally perfused with surviving cells were available for surgery. Given the technical difficulties that are today persisting in the realization of such type of grafts, the conception of softer grafts would possibly be a good compromise. In fact, the fact of realizing soft grafts (for example, hydrogels) would have some advantages such as:

– mimic the elastic properties of softer tissues which are naturally found in the first phases after bone fracture;

– allow an easier conception of internal architectures and porosities for an optimal diffusion of cells;

– allow us to equip the material with pseudo-vessels which are needed to convey the nutriments for the cells' survival.

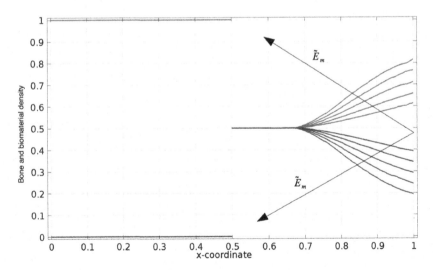

Figure 4.6. *Final distribution of natural bone and artificial biomaterial mass densities for different values of the dimensionless Young's modulus of the biomaterial $\tilde{E}_m \in [0.8, 1.2]$. For a color version of the figure see www.iste.co.uk/madeo/continuum.zip*

However, as suggested by Figure 4.6, the resorption of such material would probably be more difficult than that of a more rigid one in which the cells are equivalently distributed. Nevertheless, the use of a softer material seems today the best compromise to reach the goal of creating a final bone/biomaterial composite graft with optimal mechanical properties.

4.5. Conclusions

In this chapter, we have shown how an internal variable model (which is one of the generalized continuum models introduced in this chapter) is well adapted to describe the remodeling of living tissues in the presence of artificial bioresorbable grafts at the macroscopic scale. The introduction of two extra scalar degrees of freedom (ρ_b^* and ρ_m^*) allows us to account for the evolution of both bone and biomaterial mass densities, thus allowing for the description of the macroscopic effect of cells' activity at the lower scales.

The introduced generalized continuum model is intrinsically macroscopic in the sense that the activity of the cells acting at the lower scales is accounted for only through the evolution of the apparent mass density of both bone and biomaterial. This means that the proposed model is not able to integrate precise information about the action of single cells, but, provided that such cells are present in the considered material, the model is able to account for the overall effect of their activity at the macroscopic scale.

In the view of the design of relatively extended grafts for the treatment of important bone defects, the use of a macroscopic, continuum model is an obliged step due to the possibility of its easy implementation in a finite element code. It is clear that such a continuum model must necessarily contain some simplifying assumptions concerning the cell activity and diffusion through the material. Nevertheless, such simplifications are in some sense necessary if we want to reach a compromise between a performant macroscopic design tool and a reasonable computational time. Indeed, it is not conceivable to model macroscopic grafts of bioresorbable materials by accounting both for the activity of single cells and for the local properties of the material at small scales. It is for all these reasons that a collaborative interaction is needed at the present state of knowledge between biologists and mechanicians in order to guide the optimal design of bioresorbable grafts for the treatment of macroscopic bone defects. Such intelligent grafts should be engineered according to the following guidelines which are suggested by the results obtained in this chapter by means of the introduced generalized continuum model:

– the graft should have suitable microstructural architectures and interconnected porosity in order to allow the diffusion of cells everywhere in the graft and to get the maximum replacement of biomaterial with new-formed bone;

– the graft should be "intelligent" in the sense of being able to provide the needed nutriment to the cells, at least in the first phases after surgery while waiting for the formation of vascularization which would then assure the subsequent survival of the cells in the longer run;

– the graft should have suitable elastic properties that allow optimal strains when treated, for example, with ultrasounds. In fact, a treatment with ultrasounds corresponds in our model to the application of an external load which triggers deformations in both bone and biomaterial so rendering more efficient the resorption of the graft and its gradual replacement with new-formed bone.

Once the materials of this type will be available, the proposed generalized continuum model will provide a powerful tool for the conception of optimized bioresorbable grafts for the treatment of extended bone defects. In fact, suitable finite element code can be set up on the basis of the proposed model which allows us to optimize the graft in terms of shape, elastic properties, applied external loads, etc.

We explicitly mention the fact that, even if deal with a simple 1D case here, the biomechanical model presented in section 4.2 is indeed general, so that two-dimensional (2D) and even three-dimensional (3D) cases can be easily confronted in the same spirit of what is done here for the 1D case.

5

Energy Dissipation in Modified
and Unmodified Concrete

In this chapter, we show some results concerning the description of the dynamic behavior of modified and unmodified concrete via generalized continuum theories.

Concrete is a material which can be considered to be heterogeneous at the scale of laboratory specimens. In fact, as already remarked in the introductory chapter microcracks can be identified inside the concrete matrix that may be seen to have a macroscopic effect on the overall mechanical behavior of such an engineering material. In particular, when considering the application of cyclic loads, some characteristic hysteretic behaviors can be observed which are particular to the considered material (see Figure 5.1).

As is well known, the area of the hysteresis loop depicted in Figure 5.1 can be directly associated with the energy which is dissipated by the considered system during a loading/unloading cycle. The physical mechanism associated with this energy dissipation phenomenon can be summarized as follows: the application of a compressive load generates a slipping of the microcracks which are present inside the concrete matrix and a certain amount of energy is dissipated because of friction. Other energy is lost during the unloading phase (always by friction) when the cracks ideally return to their initial position and hence a specular slipping can be observed with respect to the loading phase. We can state that the amount of energy which is lost by friction due to microscopic relative motion of the crack lips has a macroscopic manifestation which is visible in the non-negligible area of the observed hysteresis loop. In fact, the macroscopic quantification of the energy which is lost by friction during a loading/unloading cycle can be directly obtained by estimating the area of the hysteresis loop presented in Figure 5.1 and supposing that other possible energy losses due, for example, to friction in the used machine are negligible with respect to the energy loss inside the material. In this section, we discuss how a simple internal variable model can be used to effectively predict such macroscopic manifestation of the microstructure of concrete. In particular, we will show that the fact of considering a supplementary, scalar, kinematic variable and a set of suitable constitutive relations both for the strain energy density and the dissipation functional will permit the description of dissipation phenomena observed in concrete in targeted laboratory test. Such laboratory tests are conceived in order to allow the fitting of the introduced macroscopic parameters to real experimental evidence, but that still need to be completed in order to provide a systematic identification of such coefficients.

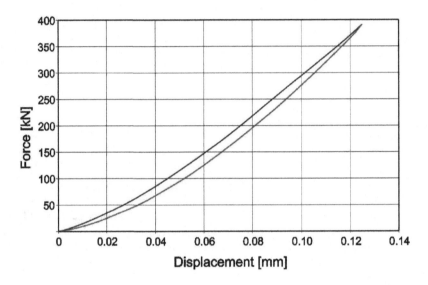

Figure 5.1. *Experimental hysteretic behavior observed in concrete subjected to cyclic load. For a color version of the figure see www.iste.co.uk/madeo/continuum.zip*

Once the material parameters are suitably identified, we can think of fitting them to specimens of concrete which have been modified with fillers that are thought to change the microscopical friction properties of the considered materials. Much research has been performed (see, for example, [MAD 06, BOW 12]) which shows that the addition of some inert fillers to the cement past can vary the quantity of energy which is dissipated by the obtained modified concrete. If the parameters of the proposed generalized continuum model are fitted to a significant number of experimental tests performed on modified and unmodified concrete, then some specific patterns of the variation of the introduced constitutive coefficients as function of the type and quantity of the used fillers could be identified. These specific patterns would then serve as a guide for the conception of optimized materials which present higher performances with respect to energy dissipation and improved (or unchanged) mechanical properties with respect to standard concrete. Such materials could be useful for applications associated with vibration control in concrete structures close to sources of vibration (train and tram stations, pavements, etc.) or, when suitably coupled with steel reinforcements, even for optimization of buildings in seismic zones.

5.1. A simple generalized continuum model for microstructure-related friction

We present in this section a simplified internal variable model which accounts for the introduction of a unique supplementary kinematical descriptor. Our final aim is to investigate the mechanism of internal dissipation due to Coulombian friction. Many authors examine the different mechanisms of internal dissipation in brittle materials such as concrete; see, for example, for more details [LOM 57] and [BHA 93]. The additional microstructural kinematic variable will be denoted by φ and is introduced to take into account the relative slipping of opposite lips of the cracks at the microscopic level. Indeed, such microscopic motion takes place when the two faces of the microcracks come in contact and slide one with respect to the other because of the externally applied load. We assume that the deformation energy of the considered medium is characterized by a volume strain energy density W which depends on the basic kinematic descriptors, i.e. the classical displacement field u and the new scalar field φ which represents the relative displacement of two superimposed lips of considered microcracks. In particular, we set:

$$W\left(\varepsilon, \varphi\right) = \frac{1}{2}\left(2\mu\,\varepsilon|\varepsilon + \lambda\left(\text{tr}\,\varepsilon\right)^2\right) + \frac{1}{2}k_1\varphi^2 + \frac{1}{3}k_2\varphi^3 + \frac{1}{4}k_3\varphi^4 + \alpha\varphi\sqrt{I_2^{(d)}} \quad [5.1]$$

where λ and μ are the Lamé parameters for linearly elastic isotropic materials, $\varepsilon = \left(\nabla u + \nabla u^T\right)/2$ is the linearized Green–Lagrange deformation tensor and the scalar $I_2^{(d)}$ is the second invariant of the deviatoric strain tensor dev $\varepsilon = \varepsilon - \frac{1}{3}\text{tr}(\varepsilon)\mathbf{I}$ defined as:

$$I_2^{(d)} = \frac{1}{2}\text{tr}\left(\text{dev}\,\varepsilon \cdot \text{dev}\,\varepsilon\right).$$

Since no space gradients of the introduced additional kinematical variable φ appear in the expression of the energy, we can classify the introduced model as an internal variable model according to the definition introduced in section 1.2. The ansatz [5.1] on the strain energy density will be justified a posteriori on the basis of available experimental evidence, but we can start noticing that it allows us to account for linear-elastic macroscopic deformations, while the microscopic motions are constitutively allowed to be nonlinear; a micro-macro coupling is also allowed by means of the elastic coefficient α. However, the kinetic energy density of the considered system is defined by:

$$T = \frac{1}{2}\rho\,\dot{\mathbf{u}}^2 + \frac{1}{2}\rho_\varphi\,\dot{\varphi}^2 \qquad [5.2]$$

where ρ is the mass density of the bulk material and ρ_φ is an effective macroscopic mass density linked to the microstructural variable φ. It can be checked that, since the microstrain field φ is assumed to have the dimensions of a length, the units of the parameter ρ_φ are the same as the units of the macroscopic bulk mass density ρ. Since the dissipation is not negligible, the governing equations of the considered medium do not have a variational structure, but they possess a quasi-variational structure (see, for example, [DEL 09b] and references there cited). In this context, we introduce a Rayleigh potential $\mathcal{R}$, which is aimed to describe Coulomb-type friction dissipation, in the form:

$$\mathcal{R} = \zeta\,\mathrm{tr}\varepsilon\left(\frac{\mathrm{Log}(\,\mathrm{Cosh}(\,\eta\,\dot{\varphi}\,)\,)}{\eta}\right), \qquad [5.3]$$

where ζ and η are the constitutive constants. We start by remarking that the quantity $\partial\mathcal{R}/\partial\dot{\varphi}$ represents the friction force associated with the introduced Rayleigh functional since it is in duality of the virtual microdisplacement $\delta\varphi$ when considering the expression [5.4] for the dissipative power. We also notice that the constant η accounts for the smoothness of the transition of the friction force from negative to positive values depending on the direction of the microsliding (see Figure 5.2), while the constant ζ proportionally accounts for dissipation due to microscopic frictional sliding.

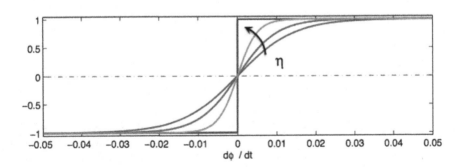

Figure 5.2. *Friction force $\partial\mathcal{R}/\partial\dot{\varphi}$ as a function of the microscopic velocity $\dot{\varphi}$. For a color version of the figure see www.iste.co.uk/madeo/continuum.zip*

Few words need to be spent here: actually, in the literature, Coulomb friction force is usually modeled by the introduction of the function signum whose argument is the velocity of the kinematical quantity on which friction forces are acting. This function, when appearing in differential equations, is a source of strong singularities and numerical or chaotic instabilities. Also basing ourselves on physical considerations, we propose to regularize the function signum with an hyperbolic tangent (see Figure 5.2) modulated with an amplitude ζ (giving the maximum of friction force which may be exerted) and with a suitably chosen slope given by η (triggering the range of velocity where friction force is an increasing function of the velocity). In conclusion, the virtual power due to internal dissipation can be written as:

$$\delta\mathcal{A}^{(Diss)}(\varepsilon, \varphi, \dot{\varphi}) = \int_0^{\bar{t}} \int_V \left(\frac{\partial \mathcal{R}}{\partial \dot{\varphi}} \delta\varphi \right) dV = \int_0^{\bar{t}} \int_V \zeta \operatorname{tr}(\varepsilon) \tanh(\eta \dot{\varphi}) \delta\varphi \, dV, \quad [5.4]$$

where V is the volume which the specimen occupies in its reference configuration and $[0, \bar{t}]$ is the time interval during which we observe the motion of the specimen. Analogously, the power of internal elastic and inertia actions as well as the power of external forces are introduced, respectively, as:

$$\delta\mathcal{A}^{(Elast)} = \delta \int_0^{\bar{t}} \int_V -W \, dV, \qquad \delta W^{(Iner)} = \delta \int_0^{\bar{t}} \int_V T \, dV,$$

$$\delta W^{(Ext)} = \int_0^{\bar{t}} \int_V \mathbf{b}^{ext} \cdot \delta\mathbf{u} \, dV,$$

with the expressions for the strain energy and kinetic energy densities given by equations [5.1] and [5.2], respectively, and where $\mathbf{b}^{ext}$ are the bulk externally applied forces (we neglect here surface external forces and microstructure-related external actions).

With the adopted notations, we can finally write the governing equations in weak form of the considered concrete-based material as:

$$\delta\mathcal{A}^{(Iner)} + \delta\mathcal{A}^{(Elast)} + \delta\mathcal{A}^{(Ext)} + \delta\mathcal{A}^{(Diss)} = 0. \qquad [5.5]$$

5.1.1. Simplified equations of motion based on Saint-Venant theory for the case of simple compression

We now suppose that the experiments which are targeted in this work are simple compressions of cylindrical concrete specimens, so that we assume that Saint-Venant theory for simple compression can be applied. In this way, we are able to deduce the simplified equations in strong form directly from expression [5.5]. To do so, we start recalling that, in the case of axial compression along the x_3 axis of a Saint-Venant cylinder, we have:

$$\varepsilon = \begin{pmatrix} -\nu\varepsilon & 0 & 0 \\ 0 & -\nu\varepsilon & 0 \\ 0 & 0 & \varepsilon \end{pmatrix}, \qquad \mathbf{u} = \begin{pmatrix} -\nu\varepsilon\,x_1 \\ -\nu\varepsilon\,x_2 \\ \varepsilon\,x_3 \end{pmatrix},$$

where we set $\varepsilon = \varepsilon_{33} = u_{3,3}$, ν is the Poisson coefficient and x_1, x_2 and x_3 are the Lagrangian coordinates (in a given reference frame with origin on the axis of the cylinder) of the material points constituting the considered specimen. We also explicitly remark that another assumption of the Saint-Venant model is that the field ε does not depend on x_1, x_2, x_3, but only, possibly, on time. We make the same assumption for the microdisplacement field φ. With these simplifying assumptions, it can be checked that in the considered particular case, integrating by parts in time, considering isochronous motions and arbitrary variations $\delta\varepsilon$ and $\delta\varphi$ the principle of virtual powers [5.5] implies the following set of ordinary differential equations in strong form:

$$\begin{cases} M\ddot{\varepsilon} + K\varepsilon + \tilde{\alpha}\varphi = f_0 + f_1\sin(\omega t) \\ m_\varphi\ddot{\varphi} + \tilde{k}_1\varphi + \tilde{k}_2\varphi^2 + \tilde{k}_3\varphi^3 + \tilde{\alpha}\varepsilon - \tilde{\zeta}\tanh(\eta\,\dot{\varphi})\,\varepsilon = 0 \end{cases} \qquad [5.6]$$

where:

$$M = \int_V \rho\left(\nu^2\left(x_1^2 + x_2^2\right) + x_3^2\right) dV, \quad m_\varphi = \int_V \rho_\varphi\, dV,$$

$$K = \int_V \left(\lambda + 2\mu + 4\left(\lambda + \mu\right)\nu^2 - 4\lambda\nu\right) dV, \quad \tilde{\alpha} = \int_V \left(\sqrt{3}/3\left(1 + \nu\right)\alpha\right) dV,$$

$$\tilde{k}_1 = \int_V k_1\, dV, \quad \tilde{k}_2 = \int_V k_2\, dV, \quad \tilde{k}_3 = \int_V k_3\, dV, \quad \tilde{\zeta} = \int_V \zeta\left(1 - 2\nu\right) dV.$$

Moreover, the external applied forces have been chosen to be those of a simple dynamic compression, i.e. $\mathbf{b}^{ext} = (0, 0, -b)^T$, with $b = b_0 + b_1\sin(\omega t)$. With this assumption, it can be checked that the loads appearing in equations [5.6] are defined as $f_0 = k\,b_0$ and $f_1 = k\,b_1$ with $k = \int_V x_3\, dV$.

Some comments are needed at this point:

– we can easily estimate the equivalent mass M by assuming that the volume mass density of concrete specimens is homogeneous (and known) while the displacement field depends linearly on reference configuration variables and on elongation;

– we assume that elastic nonlinearities are involved only in the evolution equation for the variable φ (small macroscopic strains);

– we will choose the values of the constitutive parameters in such a way that the function $\tilde{k}_1\varphi + \tilde{k}_2\varphi^2 + \tilde{k}_3\varphi^3$ is monotonously increasing;

– the deformation energy clearly needs to be definite positive and consequently suitable restrictions on the stiffness parameters must be considered;

– the amplitude of the Coulombian frictional force, for obvious physical reasons, has to be smaller than the lowest amplitude of the applied external load;

– many other possible dissipation regimes can be conceived in order to regularize the discontinuous dependence assumed in Coulomb-type model for friction: each of them would represent a different physical phenomenon, which would have different effects on the turning points of dissipation loops (i.e. when $|\dot{\varphi}|$ is suitably small). In this chapter, we assume that viscous Navier–Stokes dissipation effects are dominant when the microstructure velocity $|\dot{\varphi}|$ is suitably small, while they vanish and need to be replaced by Coulomb friction beyond a given microvelocity level.

Equations [5.6] are very similar to those presented in [BAS 00, LAM 05]. The former paper describes rheological models based on simple constitutive elements – springs, dashpot, Saint-Venant elements – and provides mathematical study, existence and uniqueness results, and an adapted numerical scheme. The latter paper considers a locally Lipschitz continuous term in the deterministic or stochastic frame.

5.2. Numerical simulations: specimen in pure compression

As far as considering numerical simulations, we limit our attention to cylindrical specimens with diameter $\phi = 11, 28$ cm and height $h = 22$ cm. All simulations are performed via the automatic code *COMSOL Multiphysics*. As for boundary conditions, the specimen is constrained at the bottom with a zero displacement in the direction x_3. A cyclic external load per unit area is applied on the top in the direction x_3 and with a frequency of 1 Hz, very low compared with the natural frequencies of the testing sample to avoid spurious inertial effects. For the same reason, an initially smooth ramp is considered as shown in Figure 5.3.

In order to illustrate the performances of the proposed model for the cement-based materials studied in this chapter, we consider several numerical simulations performed by varying the parameters introduced in equation [5.6]. If not differently specified, the constitutive parameters of the model take the values prescribed in Table 5.1. We show in Table 5.2 the ranges of values for the introduced parameters used to perform numerical parametric studies.

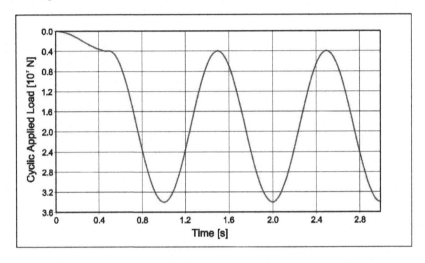

Figure 5.3. *Cyclic external load*

Parameter	Value	Unit
M	7.8×10^{-2}	kg m^2
m_φ	7.8×10^{-5}	kg
K	25×10^9	$N\,m$
$\tilde{\alpha}$	1.9×10^{11}	N

Parameter	Value	Unit
$\tilde{k}_1$	1.86×10^{12}	N/m
$\tilde{k}_2$	9.4×10^{13}	N/m^2
$\tilde{k}_3$	9.2×10^{19}	N/m^3
ζ	1.5×10^{11}	N
η	2×10^2	s/m

Table 5.1. *Reference values of the parameters used in the numerical simulations*

Parameter	Range of values	Unit
$\tilde{k}_1$	$1.7 \times 10^{12} - 1.93 \times 10^{12}$	N/m
$\tilde{k}_2$	$10^{10} - 8 \times 10^{15}$	N/m^2
$\tilde{k}_3$	$6.5 \times 10^{19} - 9.5 \times 10^{20}$	N/m^3
$\tilde{\alpha}$	$1.7 \times 10^{11} - 2 \times 10^{11}$	N
$\tilde{\zeta}$	$10^{10} - 4 \times 10^{11}$	N

Table 5.2. *Ranges of values of the material parameters used in the numerical simulations on concrete*

The numerical values shown in Table 5.1 have been determined by means of the following calibration process:

– in the static loading case, the measured stress–strain relationship has to be verified;

– the value of the parameter φ has to be compatible with the order of magnitude of known dimensions of typical cracks in concrete;

– equivalent mass coefficients are determined by taking into account volume mass density of concrete and an estimate of the percentage of total mass of concrete specimens which is moved because of crack lips movement;

– amplitude of Coulomb friction force has to be smaller than the force deforming microcracks;

– coupling between micro- and macromotion must respect definite positiveness of deformation energy.

Some parametric studies are performed in the remainder of this section on the crucial parameters of the presented model. Unless otherwise specified, the typical ranges of values assigned to the material parameters in the presented numerical simulations are those listed in Table 5.2.

5.2.1. *Effect of the basic parameters of the presented model on the area of dissipation loops*

In this section, we show the effect of any single parameter of the proposed model on the area and shape of the dissipation loops. We start by considering the effect of the Coulomb friction coefficient $\tilde{\zeta}$ on the amplitude of dissipation loops, the area of which is seen to increase monotonically with $\tilde{\zeta}$. However, because of the strong nonlinearity of the considered system, the variation of the other relevant parameters is also greatly influencing the dissipating capability as well.

Figure 5.4 shows the variation of energy dissipation loops in a stress-strain diagram when varying the friction coefficient $\tilde{\zeta}$. It can be directly remarked that the area of the loops is increased when increasing the value of the coefficient $\tilde{\zeta}$. This is completely sensible, since the mechanism which we want to associate with the parameter $\tilde{\zeta}$ is the dissipation (at the scale of the microcracks) which is due to the relative motion of the two superimposed lips of each crack as a consequence of the application of the external dynamic load. This numerical evidence is a step toward the conception of suitable experimental campaigns on concrete modified with microfillers enhancing its dissipative properties. In fact, it is sensible to assume that different microfillers with different mechanical and physical characteristics may fill the microvoids which are present inside the concrete matrix and change the microscopic friction coefficient $\tilde{\zeta}$ with respect to that of unmodified concrete.

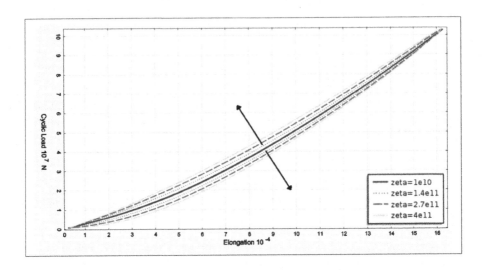

Figure 5.4. *Influence of the friction coefficient $\tilde{\zeta}$ on the energy dissipation loops for cement-based materials.* $\tilde{\zeta} = [1 \times 10^{10},\ 1.4 \times 10^{11},\ 2.7\,10^{11},\ 4 \times 10^{11}]$. *For a color version of the figure see www.iste.co.uk/madeo/continuum.zip*

Figure 5.5 shows the variation of dissipation loops when varying the coefficient $\tilde{\alpha}$, i.e. the coupling parameter between the microstructural variable φ and the macroscopic strain. It is worth noting that the effect of increasing micro-macro elastic coupling implies that the dissipation loop is shifted toward the right. Increasing the value of the coupling parameter means that the contribution to macroscopic deformation due to microscopic motion becomes increasingly important and greater macroscopic strains can be attained with the same force level. This effect allows us, for example, to relate an increasing value of the parameter $\tilde{\alpha}$ to a material with an increasing number of active cracks. Moreover, we note that the fact of increasing the coupling parameter $\tilde{\alpha}$ changes the shape of the dissipation loop giving rise to characteristic curvatures of the loading-unloading branches which are observed in available experimental curves (see, for example, Figure 5.1). We finally remark that with smaller coupling, the obtained loops can be seen to show a slightly smaller area and thus a reduced dissipation. This last effect is sensible as well since a lower number of active cracks give rise to a smaller dissipation.

To summarize, we can say that the parameter $\tilde{\alpha}$ can be seen as a quantifier of the number of cracks which are present in the considered material: the higher the parameter $\tilde{\alpha}$ (hence, the higher the number of cracks), the lower the stiffness of the material and the higher the energy dissipation.

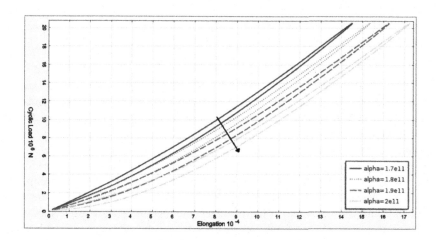

Figure 5.5. *Influence of the material parameter $\tilde{\alpha}$ on the energy dissipation loops for cement-based materials.*
$\tilde{\alpha} = [1.7 \times 10^{11},\ 1.8 \times 10^{11},\ 1.9 \times 10^{11},\ 2 \times 10^{11}]$. *For a color version of the figure see www.iste.co.uk/madeo/continuum.zip*

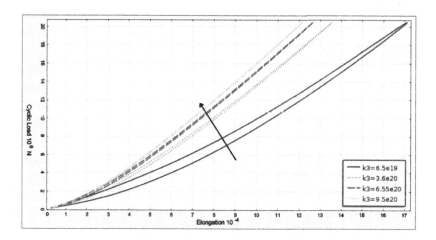

Figure 5.6. *Influence of the material parameter $\tilde{k}_3$ on dissipation loops for cement-based materials.* $\tilde{k}_3 = [6.5 \times 10^{19},\ 3.6 \times 10^{20},$
$6.55 \times 10^{20},\ 9.5 \times 10^{20}]$. *For a color version of the figure see www.iste.co.uk/madeo/continuum.zip*

In Figure 5.6, dissipation loops are depicted which show the effect of the material parameter $\tilde{k}_3$. It is possible to remark that, contrarily to what happens with the coupling parameter $\tilde{\alpha}$, high values of $\tilde{k}_3$ increase the whole stiffness of the system and consequently decrease the energy dissipated in each cycle. More precisely, $\tilde{k}_3$ can be interpreted as a microscopic stiffness which makes more difficult the relative motion of crack lips when it takes higher values (it can be interpreted as a microscopic stiffness associated with nonlinear springs connecting the two sides of considered cracks). For example, $\tilde{k}_3$ may be expected to be higher when considering a concrete matrix prepared with higher mechanical strength concrete. It can be observed (but, we do not show explicit pictures here) that if we set $\tilde{k}_3$ to be vanishing, then the characteristic sickle shape of the dissipation loop associated with Coulomb friction phenomena is completely lost and the dissipation loop resembles more an ellipsoid which is known to be associated with viscous dissipation phenomena. Hence, we can conclude that the constitutive parameter $\tilde{k}_3$, intrinsically associated with the microscopic elastic behavior governing the relative motion of crack lips, is necessary if we want to correctly describe the characteristic sickle shape of the experimental dissipation loops.

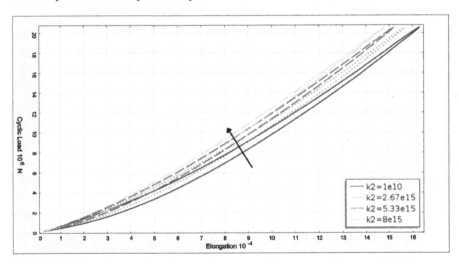

Figure 5.7. *Influence of the material parameter $\tilde{k}_2$ on energy dissipation loops for cement-based materials.*
$\tilde{k}_2 = [1 \times 10^{10}, \ 2.67 \times 10^{15}, \ 5.33 \times 10^{15}, \ 8 \times 10^{15}]$. *For a color version of the figure see www.iste.co.uk/madeo/continuum.zip*

In Figure 5.7, the behavior of dissipation loops is depicted when increasing the value of the material parameter, $\tilde{k}_2$. It can also be observed that the microstiffness parameter $\tilde{k}_2$ can be related to microscopic elastic mechanisms which makes the whole specimen more rigid when this parameter takes higher values. Nevertheless, the influence of this parameter on dissipation loop shape and amplitude is less pronounced when compared with that due to $\tilde{k}_3$. This is sensible since the microscopic nonlinearities associated with the parameter $\tilde{k}_3$ are of higher order than those associated with $\tilde{k}_2$: such parameter can be seen as the rigidity of some nonlinear microscopic springs of lower order with respect to the springs associated with $\tilde{k}_3$. Numerical simulations show that values of $\tilde{k}_2$ which span in the range $\left[0, 10^{15}\right]$ do not substantially affect the dissipation behavior of the system. However, for the values of $\tilde{k}_2$ shown in Table 5.2, a rigidifying effect of the material for higher values of $\tilde{k}_2$ can be observed.

Figure 5.8 shows how the material parameter $\tilde{k}_1$ (linear microscopic springs' stiffness) affects the stress-strain cycles. It is possible to note that an increase in the parameter $\tilde{k}_1$ does not directly affect the amount of energy dissipation, but changes the stiffness of the considered material in a similar way as $\tilde{k}_2$.

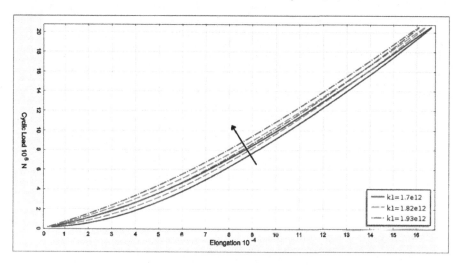

Figure 5.8. *Influence of the material parameter $\tilde{k}_1$ on energy dissipation loops for cement-based materials.*
$\tilde{k}_1 = [1.7\,10^{12},\ 1.8\,10^{12},\ 1.93\,10^{12}]$. *For a color version of the figure see www.iste.co.uk/madeo/continuum.zip*

5.2.2. *Some remarks about experimental tests and results*

Some experimental campaigns have been performed at LGCIE INSA-Lyon on modified and unmodified concrete specimens. The most important results associated with these measurement campaigns are associated with the possibility of identifying the material parameters of the proposed generalized continuum model as previously presented in Table 5.1. Nevertheless, other additional preliminary results have been obtained based on the exploitation of such measurement campaigns which allow us to draw some conclusions about the effect of the addition of calcareous fillers (see [MAD 06, BOW 12]) to the cement past.

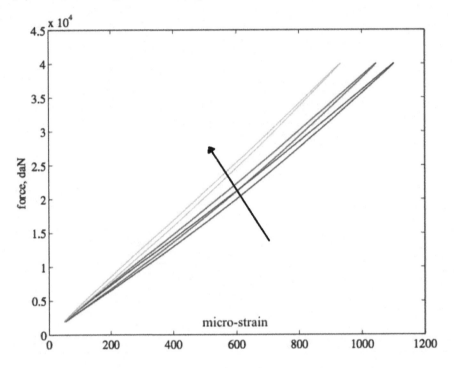

Figure 5.9. *Effect of the addition of calcareous filler GS3 on hysteresis loops for specimens with a very stiff matrix. Increasing amount of filler: from right to left. For a color version of the figure see www.iste.co.uk/madeo/continuum.zip*

We briefly list the additional indications which are furnished by such campaigns in the following points:

– When the concrete matrix is rather "soft", i.e. the compressive stress-at-failure is comparable to that of a standard concrete, then the effect of adding a calcareous filler is to increase the amount of energy dissipation. This result confirms what was previously found in [MAD 06] and is a symptom that the addition of filler actually "fills" the voids between the crack lips thus increasing the microscopic friction coefficient when a suitable relative motion of the cracks lips takes place.

– When the concrete matrix is very stiff, i.e. the compressive stress-at-failure is up to double with respect to that of a standard concrete (this effect can be obtained, for example, with high-performance cement powders), then the fact of adding a filler does not significantly change the amount of energy which is dissipated. However, when adding a larger amount of filler, the material becomes stiffer (see Figure 5.9). This fact can be explained with the fact that when we have a very stiff matrix, then the relative motion of the cracks' lips is inhibited and this fact reduces the effect of the presence of filler on energy dissipation. The addition of the filler having the effect of diminishing the void ratio in the material and being the relative motion of the crack lips inhibited, we can observe that the material becomes stiffer without significant change in the energy dissipation. Such a phenomenon is exactly the same as described before for the effect of the parameter $\tilde{\alpha}$ of our generalized model (see Figure 5.5) so that the performed measurements allow us to collect important information useful for the validation of the proposed model.

– The phenomenon of energy dissipation is strongly frequency dependent, so that such an aspect will need to be taken into account in future investigations.

It is clear that the parameter estimation plays a critical role in accurately describing the behavior of the considered mechanical system through mathematical models such as the one proposed here. Thus, subsequent studies are required to understand how good the chosen set of numerical values proposed before are (see Table 5.1). More particularly, the parameter identification previously performed needs to be more deeply investigated in future research, using, for example, fitting techniques based on the principle of "maximum likelihood" and needs suitable validation by the chi-square test to assess the reliability of the obtained fitting.

5.3. Conclusions

We have shown in this chapter how a simple internal variable model can be used for describing the dissipative behavior of concrete due to internal frictional sliding of the microcracks. The numerical results obtained show that generalized models of this type can be of use for a simplified description of the mechanical behavior of concrete when subjected to external dynamical loadings. The numerical simulations performed allowed us to obtain a first fitting of the introduced constitutive parameters by means of comparisons with the available experimental results. The main strong points of the proposed macroscopic modeling can be summarized as follows:

– the essential macroscopic characteristics of the observed material behavior are caught by introducing a few constitutive parameters;

– Coulomb-type dissipation at the microlevel is accounted for by introducing a simple Rayleigh functional;

– the introduction of a unique scalar internal variable allows for the description of microscopic motions between the superimposed faces of the internal microcracks.

Notwithstanding the potentialities of the proposed modeling, future work should investigate different generalizations in order to treat some delicate points such as:

– relating more precisely the constitutive parameters to the characteristics of the microstructures;

– further generalizing the proposed kinematics with non-scalar internal variables in order to account for possible non-uniform distributions of cracks inside the material;

– performing a significative number of experimental tests in order to have a more reliable quantitative estimate of the introduced parameters.

Bibliography

[ALT 10] ALTENBACH H., EREMEYEV V., LEBEDEV L. *et al.*, "Acceleration waves and ellipticity in thermoelastic micropolar media", *Arch. Appl. Mech.*, vol. 80, no. 3, pp. 217–227, 2010.

[ALT 13] ALTENBACH H., EREMEYEV V.A. (eds), *Generalized Continua from the Theory to Engineering Applications*, Springer, Vienna, 2013.

[AUF 13] AUFFRAY N., DELL'ISOLA F., EREMEYEV V. *et al.*, "Analytical continuum mechanics á la Hamilton–Piola least action principle for second gradient continua and capillary fluids", *Math. Mech. Solids*, 2013.

[ASK 11] ASKES H., AIFANTIS E.C., "Gradient elasticity in statics and dynamics: an overview of formulations, length scale identification procedures, finite element implementations and new results", *Int. J. Solids Struct.*, vol. 48, pp. 1962–1990, 2011.

[ALI 03] ALIBERT J.-J., SEPPECHER P., DELL'ISOLA F., "Truss modular beams with deformation energy depending on higher displacement gradients", *Math. Mech. Solids*, vol. 8, no. 1, pp. 51–73, 2003.

[AND 86] ANDRIEUX S., BAMBERGER Y., MARIGO J.J., "Un modèle de matériau microfissuré pour les roches et les bétons", *Journal de mécanique théorique et appliquée*, vol. 5, nos. 471–513, pp. 471–513, 1986.

[AND 04] ANDREAUS U., DELL'ISOLA F., PORFIRI M., "Piezoelectric passive distributed controllers for beam flexural vibrations", *Journal of Vibration and Control*, vol. 10, no. 5, pp. 625–659, 2004.

[BAS 00] BASTIEN J., SCHATZMAN M., LAMARQUE C.-H., "Study of some rheological models with a finite number of degrees of freedom", *European Journal of Mechanics – A/Solids*, vol. 19, no. 2, pp. 277–307, 2000.

[BER 09] BEREZOVSKI A., ENGELBRECHT J., MAUGIN G.A., "One-dimensional microstructure dynamics", *Mechanics of Microstructured Solids. Lecture Notes in Applied and Computational Mechanics*, vol. 46, pp. 21–28, 2009.

[BHA 93] BHATTACHARJEE S.S., LÉGER P., "Seismic cracking and energy dissipation in concrete gravity dams", *Earthquake Engineering and Structural Dynamics*, John Wiley & Sons, Ltd, vol. 22, no. 11, pp. 991–1007, 1993.

[BIT 92] BITTNER L., DACOROGNA B., "Direct methods in the calculus of variations. Berlin etc., Springer-Verlag, 1989. IX, p. 308, DM 120, (Applied Mathematical Sciences 78)", *Zeitschrift für Angewandte Mathematik und Mechanik (ZAMM)*, vol. 72, no. 3, p. 239, Wiley-VCH Verlag, 1992.

[BLE 67] BLEUSTEIN J.L., "A note on the boundary conditions of Toupin's strain gradient-theory", *Int. J. Solids Structures*, vol. 3, pp. 1053–1057, 1967.

[BOU 13] BOUTIN C., SOUBESTRE J., DIETZ M.S. *et al.*, "Experimental evidence of the high-gradient behaviour of fiber reinforced materials", *European Journal of Mechanics – A/Solids*, vol. 42, pp. 280–298, 2013.

[BOW 12] BOWLAND A.G., WEYERS R.E., CHARNEY F.A. *et al.*, "Effect of vibration amplitude on concrete with damping additives", *ACI Materials Journal*, vol. 109, no. 3, pp. 371–378, 2012.

[BUE 03] BUECHNER P.M., LAKES R.S., "Size effects in the elasticity and viscoelasticity of bone", *Biomechanics and Modeling in Mechanobiology*, Springer-Verlag, vol. 1, no. 4, pp. 295–301, 2003.

[CAR 96] CARTER D.R., VAN DER MEULEN M.C.H., BEAUPRE G.S., "Mechanical factors in bone growth and development", *Bone*, Elsevier, vol. 18, no. 1, pp. S5–S10, 1996.

[CAS 10] CASANOVA R., MOUKOKO D., PITHIOUX M. *et al.*, "Temporal evolution of skeletal regenerated tissue: what can mechanical investigation add to biological?", *Med. Biol. Eng. Comput.*, vol. 48, pp. 811–819, 2010.

[CHA 11] CHARMETANT A., Approches hyperélastiques pour la modélisation du comportement mécanique de préformes tissées de composites, PhD Thesis, INSA-Lyon, 2011.

[CHA 12] CHARMETANT A., ORLIAC J.G., VIDAL SALLÉE E. *et al.*, "Hyperelastic model for large deformation analyses of 3D interlock composite preforms", *Composites Science and Technology*, vol. 72, pp. 1352–1360, 2012.

[CHE 12] CHESNAIS C., BOUTIN C., HANS S., "Effects of the local resonance on the wave propagation in periodic frame structures: generalized Newtonian mechanics", *The Journal of the Acoustical Society of America*, vol. 132, no. 4, pp. 2873–2886, 2012.

[COS 09] COSSERAT E., COSSERAT F., *Théorie des corps déformables*, Hermann, Paris, France, 1909.

[COW 76] COWIN S.C., HEGEDUS D.H., "Bone remodeling I: theory of adaptive elasticity", *Journal of Elasticity*, Kluwer Academic Publishers, vol. 6, no. 3, pp. 313–326, 1976.

[COW 01] COWIN S., *Bone Mechanics Handbook, 2nd ed.*, CRC Press, 2001.

[DE 81] DE GENNES P.G., "Some effects of long range forces on interfacial phenomena", *Journal de Physique Lettres*, Les Editions de Physique, vol. 42, no. 16, pp. 377–379, 1981.

[DEL 98] DELL'ISOLA F., VIDOLI S., "Continuum modelling of piezoelectromechanical truss beams: an application to vibration damping", *Archive of Applied Mechanics*, vol. 68, no. 1, pp. 1–19, 1998.

[DEL 09a] DELL'ISOLA F., SCIARRA G., VIDOLI S., "Generalized Hooke's law for isotropic second gradient materials", *Proceedings of the Royal Society A: Mathematical, Physical and Engineering Science*, vol. 465, no. 2107, pp. 2177–2196, 2009.

[DEL 09b] DELL'ISOLA F., MADEO A., SEPPECHER P., "Boundary conditions at fluid-permeable interfaces in porous media: a variational approach", *International Journal of Solids and Structures*, vol. 46, no. 17, pp. 3150–3164, 2009.

[DEL 09c] DE LUYCKER E., MORESTIN F., BOISSE P. *et al.*, "Simulation of 3D interlock composite preforming", *Composite Structures*, Elsevier, vol. 88, no. 4, pp. 615–623, 2009.

[DEL 14a] DELL'ISOLA F., STEIGMANN D., "A two-dimensional gradient-elasticity theory for woven fabrics", *Journal of Elasticity*, 2014.

[DEL 14b] DELL'ISOLA F., ANDREAUS U., PLACIDI L., "At the origins and in the vanguard of peridynamics, non-local and higher-gradient continuum mechanics: an underestimated and still topical contribution of Gabrio Piola", *Mathematics and Mechanics of Solids*, Sage Publications, 2014.

[DEL 14c] DELL'ISOLA F., MAIER G., PEREGO U., *The Complete Works of Gabrio Piola*, of *Advanced Structured Materials*, Springer, New York, vol. 38, 2014.

[DUM 87] DUMONT J., LADEVÈZE P., POSS M. *et al.*, "Damage mechanics for 3-D composites", *Composite Structures*, Elsevier, vol. 8, no. 2, pp. 119–141, 1987.

[DOB 02] DOBLARÉ M., GARCÍA J.M., "Anisotropic bone remodelling model based on a continuum damage-repair theory", *Journal of Biomechanics*, Elsevier, vol. 35, no. 1, pp. 1–17, 2002.

[ECO 94] ECONOMOAU E.N., SIGALABS M., "Stop bands for elastic waves in periodic composite materials", *J. Acoust. Soc. Am.*, vol. 95, no. 4, pp. 1734–1740, 1994.

[ERE 14] EREMEYEV V., ALTENBACH H., "Equilibrium of a second-gradient fluid and an elastic solid with surface stresses", *Meccanica*, vol. 49, no. 11, pp. 2635–2643, 2014.

[ERI 99] ERINGEN A.C., *Microcontinuum Field Theories I. Foundations and Solids*, Springer-Verlag, New York, 1999.

[EXA 01] EXADAKTYLOS G.E., VARDOULAKIS I., "Microstructure in linear elasticity and scale effects a reconsideration of basic rock mechanics and rock fracture mechanics", *Tectonophysics*, vol. 335, pp. 81–109, 2001.

[ENG 06] ENGHETA N., ZIOLKOWSKI R.W., *Metamaterials: Physics and Engineering Explorations*, Wiley, New York, 2006.

[FER 14] FERRETTI M., MADEO A., DELL'ISOLA F. *et al.*, "Modelling the onset of shear boundary layers in fibrous composite reinforcements by second gradient theory", *ZAMP*, vol. 65, no. 3, pp. 587–612, 2014.

[FOR 98] FOREST S., "Mechanics of generalized continua: construction by homogenization", *Journal de Physique IV*, vol. 8, pp. Pr8-39–48, 1998.

[FOR 99a] FOREST S., "Homogenization methods and the mechanics of generalized continua", *Geometry, Continua and Microstructure*, Hermann, Paris, France, pp. 35–48, 1999.

[FOR 99b] FOREST S., DENDIEVEL R., CANOVA G., "Estimating the overall properties of heterogeneous Cosserat materials", *Modelling and Simulation in Materials Science and Engineering*, IOP Publishing, vol. 7, no. 5, p. 829, 1999.

[FOR 01] FOREST S., PRADEL F., SAB K., "Asymptotic analysis of heterogeneous Cosserat media", *International Journal of Solids and Structures*, vol. 38, no. 2627, pp. 4585–4608, 2001.

[FOR 02] FOREST S., "Homogenization methods and the mechanics of generalized continua – part 2", *Theoretical and Applied Mechanics*, vol. 28–29, pp. 113–148, 2002.

[FOR 10] FOREST S., AIFANTIS E.C., "Some links between recent gradient thermo-elasto-plasticity theories and the thermomechanics of generalized continua", *International Journal of Solids and Structures*, vol. 47, no. 2526, pp. 3367–3376, 2010.

[FRA 86] FRANCFORT G., SUQUET P., "Homogenization and mechanical dissipation in thermoviscoelasticity", *Archive for Rational Mechanics and Analysis*, Springer-Verlag, vol. 96, no. 3, pp. 265–293, 1986.

[GER 73a] GERMAIN P., "La méthode des puissances virtuelles en mécanique des milieux continus-I: Théorie du second gradient", *J. Mécanique*, vol. 12, pp. 235–274, 1973.

[GER 73b] GERMAIN P., "The method of virtual power in continuum mechanics. Part 2: microstructure", *SIAM J. Appl. Math.*, vol. 25, pp. 556–575, 1973.

[GHI 13] GHIBA I.D., NEFF P., MADEO A. *et al.*, "The relaxed linear micromorphic continuum: existence, uniqueness and continuous dependence in dynamics", *Submitted to Mathematics and Mechanics of Solids*, 2013.

[GÓM 11] GÓMEZ-BENITO M.J., GONZÁLEZ-TORRES L.A., REINA-ROMO E. *et al.*, "Influence of high-frequency cyclical stimulation on the bone fracture-healing process: mathematical and experimental models", *Philosophical Transactions of the Royal Society of London A: Mathematical, Physical and Engineering Sciences*, The Royal Society, vol. 369, no. 1954, pp. 4278–4294, 2011.

[GON 10] GONZÁLEZ-TORRES L., GÓMEZ-BENITO M., DOBLARÉ M. *et al.*, "Influence of the frequency of the external mechanical stimulus on bone healing: a computational study", *Medical Engineering and Physics*, Elsevier, vol. 32, no. 4, pp. 363–371, 2010.

[GRÜ 88] GRÜNDEMANN H., "Homogenization techniques for composite media" in SANCHEZ-PALENCIA E., ZAOUI A. (eds), Proceedings, Udine, Italy, 1985. Berlin etc., Springer-Verlag, 1987. IX, p. 397, DM 73, (Lecture Notes in Physics 272)", *Zeitschrift für Angewandte Mathematik und Mechanik (ZAMM)*, vol. 68, no. 6, pp. 212–212, Wiley-VCH Verlag, 1988.

[HAS 10] HASUIKE A. *et al.*, "In vivo bone regenerative effect of low-intensity pulsed ultrasound in rat calvarial defects", *Oral Surg. Oral Med. Oral Pathol. Oral Radiol. Endod.*, vol. 111, pp. e12–e20, 2010.

[HAM 13a] HAMILA N., BOISSE P., "Locking in simulation of composite reinforcement deformations. Analysis and treatment", *Composites Part A: Applied Science and Manufacturing*, Elsevier, vol. 53, pp. 109–117, 2013.

[HAM 13b] HAMILA N., BOISSE P., "Tension locking in finite-element analyses of textile composite reinforcement deformation", *Comptes Rendus Mécanique*, vol. 341, no. 6, pp. 508–519, 2013.

[HAR 12] HARRISON P., "Normalisation of biaxial bias extension test results considering shear tension coupling", *Composites Part A: Applied Science and Manufacturing*, vol. 43, no. 9, pp. 1546–1554, September 2012.

[HOC 06] HOC T., HENRY L., VERDIER M. *et al.*, "Effect of microstructure on the mechanical properties of Haversian cortical bone", *Bone*, vol. 38, no. 4, pp. 466–474, 2006.

[HUI 00] HUISKES R., RUIMERMAN R., VAN LENTHE G.H. *et al.*, "Effects of mechanical forces on maintenance and adaptation of form in trabecular bone", *Nature*, vol. 405, pp. 704–706, 2000.

[JEO 09] JEONG J., RAMEZANI H., MÜNCH I. *et al.*, "A numerical study for linear isotropic Cosserat elasticity with conformally invariant curvature", *Z. Angew. Math. Mech.*, vol. 89, no. 7, pp. 552–569, 2009.

[JEO 10] JEONG J., NEFF P., "Existence, uniqueness and stability in linear Cosserat elasticity for weakest curvature conditions", *Math. Mech. Solids*, vol. 15, no. 1, pp. 78–95, 2010.

[KAF 00] KAFESAKI M., SIGALAS M.M., GARCÍA N., "Frequency modulation in the transmittivity of wave guides in elastic-wave band-gap materials", *Physical Review Letters*, vol. 85, no. 19, pp. 4044–4047, 2000.

[KRU 98] KRUCH S., FOREST S., "Computation of coarse grain structures using a homogeneous equivalent medium", *Journal de Physique IV*, vol. 8, pp. Pr8-197–205, 1998.

[LAD 85] LADEVÈZE P., PROSLIER L., REMOND Y., "Reconstruction of a 3-D composite behaviour from a local approach", *ICCM-V*, pp. 1025–1037, 1985.

[LAK 82] LAKES R.S., "Dynamical study of couple stress effects in human compact bone", *Journal of Biomechanical Engineering*, vol. 104, no. 1, pp. 6–11, 1982.

[LAM 05] LAMARQUE C.-H., BERNARDIN F., BASTIEN J., "Study of a rheological model with a friction term and a cubic term: deterministic and stochastic cases", *European Journal of Mechanics – A/Solids*, vol. 24, no. 4, pp. 572–592, 2005.

[LAV 09] LAVANDIER B., GLEIZAL A., BÉRA J.-C., "Experimental assessment of calvarial bone defect re-ossification stimulation using low-intensity pulsed ultrasound", *Ultrasound in Med. and Biol.*, vol. 35, no. 4, pp. 585–594, 2009.

[LEE 08] LEE W., PADVOISKIS J., CAO J. *et al.*, "Bias-extension of woven composite fabrics", *International Journal of Material Forming*, Springer-Verlag, vol. 1, no. 1, pp. 895–898, 2008.

[LIU 00] LIU Z., ZHANG X., MAO Y. *et al.*, "Locally resonant sonic materials", *Science*, American Association for the Advancement of Science, vol. 289, no. 5485, pp. 1734–1736, 2000.

[LOI 08] LOIX F., BADEL P., ORGÉAS L. *et al.*, "Woven fabric permeability: from textile deformation to fluid flow mesoscale simulations", *Composites Science and Technology*, Elsevier, vol. 68, no. 7, pp. 1624–1630, 2008.

[LOM 57] LOMNITZ C, "Linear dissipation in solids", *Journal of Applied Physics*, vol. 28, no. 2, pp. 201–205, 1957.

[MAD 06] MADEO A., Effect of micro-particle additions on frictional energy dissipation and strength of concrete, PhD Thesis, Virginia Polytechnic Institute and State University, 2006.

[MAD 11] MADEO A., LEKSZYCKI T., DELL'ISOLA F., "A continuum model for the bio-mechanical interactions between living tissue and bio-resorbable graft after bone reconstructive surgery", *CRAS Mécanique*, vol. 339, no. 10, pp. 625–640, 2011.

[MAD 12] MADEO A., GEORGE D., LEKSZYCKI T. *et al.*, "A second gradient continuum model accounting for some effects of micro-structure on reconstructed bone remodelling", *Comptes Rendus Mécanique*, vol. 340, no. 8, pp. 575–589, 2012.

[MAD 13] MADEO A., NEFF P., GHIBA I.D. *et al.*, "Wave propagation in relaxed micromorphic continua: modeling metamaterials with frequency band-gaps", *Continuum Mechanics and Thermodynamics*, 2013.

[MAD 14a] MADEO A., PLACIDI L., ROSI G., "Towards the design of metamaterials with enhanced damage sensitivity: second gradient porous materials", *Research in Nondestructive Evaluation*, vol. 25, no. 2, pp. 99–124, 2014.

[MAD 14b] MADEO A., NEFF P., GHIBA I.-D. *et al.*, "Band gaps in the relaxed linear micromorphic continuum", *Z. Angew. Math. Mech. (ZAMM)*, June 2014.

[MAD 15] MADEO A., GHIBA I.-D., NEFF P. *et al.*, "Incomplete traction boundary conditions in the Grioli-Koiter-Mindlin-Toupin indeterminate couple stress model", submitted 2015.

[MAR 85] MARIGO J.J., "Modelling of brittle and fatigue damage for elastic material by growth of microvoids", *Engineering Fracture Mechanics*, vol. 21, no. 4, pp. 861–874, 1985.

[MAN 13] MAN W., FLORESCU M., MATSUYAMA K. *et al.*, "Photonic band gap in isotropic hyperuniform disordered solids with low dielectric contrast", *Opt. Express*, OSA, vol. 21, no. 17, pp. 19972–19981, August 2013.

[MAU 99] MAUGIN G., *Nonlinear Waves in Elastic Crystals*, Oxford University Press, 1999.

[MAU 04] MAURINI C., DELL'ISOLA F., POUGET J., "On models of layered piezoelectric beams for passive vibration control", *Journal De Physique*, vol. 115, pp. 307–316, 2004.

[MAU 06] MAURINI C., POUGET J., DELL'ISOLA F., "Extension of the Euler-Bernoulli model of piezoelectric laminates to include 3D effects via a mixed approach", *Computers and Structures*, vol. 84, nos. 22–23, pp. 1438–1458, 2006.

[MER 10a] MERKEL A., TOURNAT V., "Dispersion of elastic waves in three-dimensional noncohesive granular phononic crystals: properties of rotational modes", *Physical Review E*, vol. 82, p. 031305, 2010.

[MER 10b] MERKEL A., TOURNAT V., GUSEV V., "Elastic waves in noncohesive frictionless granular crystals", *Ultrasonics*, vol. 50, pp. 133–138, 2010.

[MER 11] MERKEL A., TOURNAT V., "Experimental evidence of rotational elastic waves in granular phononic crystals", *Physical Review Letters*, vol. 107, p. 225502, 2011.

[MIN 64] MINDLIN R.D., "Micro-structure in linear elasticity", *Arch. Rat. Mech. Analysis*, vol. 16, no. 1, pp. 51–78, 1964.

[MIN 65] MINDLIN R.D., "Second gradient of strain and surface tension in linear elasticity", *Int. J. Solids Struct.*, vol. 1, pp. 417–438, 1965.

[MIN 68] MINDLIN R.D., ESHEL N.N., "On first strain-gradient theories in linear elasticity", *Int. J. Solids Struct.*, vol. 4, pp. 109–124, 1968.

[MUL 94] MULLENDER M., HUISKES R., WEINANS H., "A physiological approach to the simulation of bone remodeling as a self-organizational control process", *J. Biomech.*, vol. 27, no. 11, pp. 1389–1394, 1994.

[NEF 13] NEFF P., GHIBA I.D., MADEO A. *et al.*, "A unifying perspective: the relaxed linear micromorphic continuum", *Submitted to Continuum Mechanics and Thermodynamics*, 2013.

[NOM 00] NOMURA S., TAKANO-YAMAMOTO T., "Molecular events caused by mechanical stress in bone", *Matrix Biology*, vol. 19, pp. 91–96, 2000.

[NEF 06] NEFF P., "The Cosserat couple modulus for continuous solids is zero viz the linearized Cauchy-stress tensor is symmetric", *Z. Angew. Math. Mech.*, vol. 86, pp. 892–912, 2006.

[NEF 07] NEFF P., FOREST S., "A geometrically exact micromorphic model for elastic metallic foams accounting for affine microstructure. Modelling, existence of minimizers, identification of moduli and computational results", *J. Elasticity*, vol. 87, pp. 239–276, 2007.

[NEF 09] NEFF P., JEONG J., "A new paradigm: the linear isotropic Cosserat model with conformally invariant curvature energy", *Z. Angew. Math. Mech.*, vol. 89, no. 2, pp. 107–122, 2009.

[NEF 10] NEFF P., JEONG J., FISCHLE A., "Stable identification of linear isotropic Cosserat parameters: bounded stiffness in bending and torsion implies conformal invariance of curvature", *Acta Mechanica*, vol. 211, nos. 3–4, pp. 237–249, 2010.

[ORL 12] ORLIAC J.G., Analyse et simulation du comportement anisotrope lors de la mise en forme de renforts tissés interlock., PhD Thesis, INSA-Lyon, 2012.

[OUI 03] OUISSE M., GUYADER J.L., "Vibration sensitive behaviour of a connecting angle. Case of coupled beams and plates", *Journal of Sound and Vibration*, vol. 267, no. 4, pp. 809–850, 2003.

[PEN 01] PENSÉE V., KONDO D, "Une analyse micromécanique 3-D de l'endommagement par mésofissuration", *Comptes Rendus de l'Académie des Sciences – Series {IIB} – Mechanics*, vol. 329, no. 4, pp. 271–276, 2001.

[PID 97] PIDERI C., SEPPECHER P., "A second gradient material resulting from the homogenization of an heterogeneous linear elastic medium", *Contin. Mech. Thermodyn.*, vol. 9, pp. 241–257, 1997.

[PIE 09a] PIETRASZKIEWICZ W., EREMEYEV V., "On natural strain measures of the non-linear micropolar continuum", *Int. J. Solids Struct.*, vol. 46, pp. 774–787, 2009.

[PIE 09b] PIETRASZKIEWICZ W., EREMEYEV V., "On vectorially parameterized natural strain measures of the non-linear Cosserat continuum", *Int. J. Solids Struct.*, vol. 46, pp. 2477–2480, 2009.

[PLA 13] PLACIDI L., ROSI G., GIORGIO I. *et al.*, "Reflection and transmission of plane waves at surfaces carrying material properties and embedded in second-gradient materials", *Mathematics and Mechanics of Solids*, 2013.

[PLA 14] PLACIDI L., "A variational approach for a nonlinear 1-dimensional second gradient continuum damage model", *Continuum Mechanics and Thermodynamics*, Springer Berlin Heidelberg, pp. 1–16, 2014.

[POR 05] PORFIRI M., DELL'ISOLA F., SANTINI E., "Modeling and design of passive electric networks interconnecting piezoelectric transducers for distributed vibration control", *International Journal of Applied Electromagnetics and Mechanics*, vol. 21, no. 2, pp. 69–87, 2005.

[PRE 94] PRENDERGAST P.J., TAYLOR D., "Prediction of bone adaptation using damage accumulation", *Journal of Biomechanics*, Elsevier, vol. 27, no. 8, pp. 1067–1076, 1994.

[RAO 09] RAOULT A., "Symmetry groups in nonlinear elasticity: an exercise in vintage mathematics", *Communications on Pure and Applied Analysis*, vol. 8, no. 1, pp. 435–456, 2009.

[RIN 14] RINALDI A., PLACIDI L., "A microscale second gradient approximation of the damage parameter of quasi-brittle heterogeneous lattices", *Z. Angew. Math. Mech. (ZAMM)*, vol. 94, no. 10, pp. 862–877, 2014.

[ROS 13] ROSI G., GIORGIO I., EREMEYEV V., "Propagation of linear compression waves through plane interfacial layers and mass adsorption in second gradient fluids", *Z. Angew. Math. Mech. (ZAMM)*, 2013.

[RUI 05] RUIMERMAN R., HILBERS P., VAN RIETBERGEN B. *et al.*, "A theoretical framework for strain-related trabecular bone maintenance and adaptation", *Journal of Biomechanics*, vol. 38, pp. 931–941, 2005.

[SCI 08] SCIARRA G., DELL'ISOLA F., IANIRO N. *et al.*, "A variational deduction of second gradient poroelasticity I: general theory", *Journal of Mechanics of Materials and Structures*, vol. 3, no. 3, pp. 507–526, 2008.

[SCH 05] SCHRDER J., NEFF P., BALZANI D., "A variational approach for materially stable anisotropic hyperelasticity", *International Journal of Solids and Structures*, vol. 42, no. 15, pp. 4352–4371, 2005.

[SEP 11] SEPPECHER P., ALIBERT J.-J., DELL'ISOLA F., "Linear elastic trusses leading to continua with exotic mechanical interactions", *Journal of Physics: Conference Series*, vol. 319, no. 1, p. 012018, 2011.

[SPE 84] SPENCER A.J.M., "Constitutive theory for strongly anisotropic solids", A.J.M. SPENCER (ed.), *Continuum Theory of the Mechanics of Fibre-Reinforced Composites*, of *International Centre for Mechanical Sciences*, vol. 282, pp. 1–32, Springer, Vienna, 1984.

[TOU 62] TOUPIN R.A., "Elastic materials with couple stresses", *Arch. Rat. Mech. Anal.*, vol. 11, pp. 385–413, 1962.

[TOU 64] TOUPIN R.A., "Theory of elasticity with couple stresses", *Arch. Rat. Mech. Anal.*, vol. 17, pp. 85–112, 1964.

[VAS 98] VASSEUR J.O., DEYMIER P.A. *et al.*, "Experimental evidence for the existence of absolute acoustic band gaps in two-dimensional periodic composite media", *J. Phys. Condens. Matter*, vol. 10, p. 6051, 1998.

[VAS 01] VASSEUR J.O., DEYMIER P.A. *et al.*, "Experimental and theoretical evidence for the existence of absolute acoustic band gaps in two-dimensional solid phononic crystals", *Physical Review Letters,*, vol. 86, no. 14, pp. 3012–3015, 2001.

[VID 01] VIDOLI S., DELL'ISOLA F., "Vibration control in plates by uniformly distributed PZT actuators interconnected via electric networks", *European Journal of Mechanics, A/Solids*, vol. 20, no. 3, pp. 435–456, 2001.

[WEI 92] WEINANS H., HUISKES R., GROOTENBOER H.J., "The behavior of adaptive bone-remodeling simulation models", *Journal of Biomechanics*, vol. 25, no. 12, pp. 1425–1441, 1992.

[WOL 86] WOLFF J., MAQUET P., FURLONG R., *The Law of Bone Remodelling*, Springer-Verlag, 1986.

[YAN 81] YANG J.F.C., LAKES R.S., "Transient study of couple stress effects in compact bone: torsion", *Journal of Biomechanical Engineering*, vol. 103, no. 4, pp. 275–279, 1981.

[YAN 82] YANG J.F.C., LAKES R.S., "Experimental study of micropolar and couple stress elasticity in compact bone in bending", *Journal of Biomechanics*, vol. 15, no. 2, pp. 91–98, 1982.

[YAN 10] YANG Y., MISRA A., "Higher-order stress-strain theory for damage modeling implemented in an element-free Galerkin formulation", *CMES-Computer Modeling in Engineering & Sciences*, vol. 64, no. 1, pp. 1–36, 2010.

[ZHU 08] ZHU Q.Z., KONDO D., SHAO J.F., "Micromechanical analysis of coupling between anisotropic damage and friction in quasi brittle materials: role of the homogenization scheme", *International Journal of Solids and Structures*, vol. 45, no. 5, pp. 1385–1405, 2008.

[ZOU 09] ZOUHDI S., SIHVOLA A., VINOGRADOV A.P., *Metamaterials and Plasmonics: Fundamentals, Modelling Applications*, Series B: Physics and Biophysics, Springer, New York, 2009.

[ZHO 09] ZHOU X., HU G., "Analytic model of elastic metamaterials with local resonances", *Physical Review B*, APS, vol. 79, no. 19, p. 195109, 2009.

Index